POSITIONING IN WIRELESS COMMUNICATIONS SYSTEMS

POSITIONING IN WIRELESS COMMUNICATIONS SYSTEMS

Stephan Sand, Armin Dammann and Christian Mensing

German Aerospace Center (DLR), Germany

This edition first published 2014
© 2014 John Wiley & Sons, Ltd

Registered office
John Wiley & Sons Ltd, The Atrium, Southern Gate, Chichester, West Sussex, PO19 8SQ, United Kingdom

For details of our global editorial offices, for customer services and for information about how to apply for permission to reuse the copyright material in this book please see our website at www.wiley.com.

Library of Congress Cataloguing-in-publication data has been applied for

ISBN: 978-0-4707-7064-1 (hardback)

A catalogue record for this book is available from the British Library.

ISBN: 9780470770641

Typeset in 10/12pt TimesLTStd by Laserwords Private Limited, Chennai, India

1   2014

Contents

About the Authors

Stephan Sand (MSc EE 2001, Dipl-Ing 2002, Dr ETH Zurich 2010) studied electrical engineering with focus on communications technology, digital signal processing, and wireless communications at the University of Ulm, Germany (1996–2002), the University of Massachusetts Dartmouth, MA, USA (1999–2001), and the Swiss Federal Institute of Technology (ETH) Zurich, Switzerland (2005–2009). In 2002, he joined the Institute of Communications and Navigation, German Aerospace Center (DLR), Oberpfaffenhofen, Germany. Currently, he is managing and working on cooperative positioning and swarm navigation research projects at DLR. He was visiting researcher at NTT DoCoMo R&D Yokosuka, Japan in 2004 and at the Swiss Federal Institute of Technology (ETH) Zurich, Switzerland in 2007 working in the area of wireless communications. Stephan was actively involved in several research projects on mobile radio funded by the European Commission (4MORE, NEWCOM, COST289, PLUTO) and by international industry cooperation. In the GJU/GSA project GREAT and the EU FP7-ICT collaborative project WHERE, he lead the work on hybrid location determination. He was the coordinator of the recent EU FP7 project GRAMMAR on Galileo mass-market receivers. Currently, he leads the work on cooperative swarm navigation in the Valles Marineris Explorer Project. Stephan is a founding member of the International Conference on Localization and GNSS (ICL-GNSS) steering committee and was program committee co-chair of ICL-GNSS 2011 and general chair of ICL-GNSS 2012. His research interests and activities include wireless communications, multi-sensor navigation, cooperative positioning, and swarm navigation. Stephan has authored and coauthored more than 90 technical and scientific publications in conferences and journals, and obtained several patents on his inventions.

Armin Dammann (Dipl-Ing 1997, Dr-Ing 2005) studied electrical engineering with the main topic information- and microwave-technology at the University of Ulm, Germany from 1991–1997. In 1997, Armin joined the Institute of Communications and Navigation of the German Aerospace Center (DLR). Since 2005 he has been head of the Mobile Radio Transmission Research Group. His research interests and activities include navigation signal design for the European satellite navigation system Galileo, PHY/MAC layer design for terrestrial communications systems based on OFDM, antenna diversity techniques for wireless communications/broadcast systems and synchronization/positioning in wireless communications. Armin has authored and co-authored more than 120 technical and scientific publications in conference proceedings and journals in the fields of wireless communications and (cooperative) positioning. In these areas, he additionally holds several international patents. He has coorganized and cochaired the MC-SS workshop series on multi-carrier systems and solutions. Armin has been active in several EU research projects, for example, MCP, 4MORE, WINNER, NEWCOM, PLUTO, GREAT, GRAMMAR, and WHERE/WHERE2. For the latter, he has also been involved in management and coordination.

Christian Mensing (BSc 2002, Dipl-Ing 2004, MSc 2005 and Dr-Ing 2013) studied electrical engineering and information technology focusing on signal processing and high frequency technology at the Munich University of Technology (TUM), Germany. He received the BSc, Dipl-Ing, MSc, and Dr-Ing degree from TUM in 2002, 2004, 2005, and 2013, respectively. In 2005, he joined the Institute of Communications and Navigation of the German Aerospace Center (DLR) as a research engineer. His main interests included location estimation strategies for cellular networks and satellite based navigation systems, and detection techniques for communications. He has authored and co-authored more than 40 publications in conference proceedings and journals, and holds several patents. Christian has been involved in various European research projects on positioning and communications, for example, GREAT, WINNER, GRAMMAR, and WHERE. Since 2011, he has been working as development engineer at Rohde & Schwarz, Germany.

Preface

Since the advent of smartphones and tablet computers, such as Apple's iPhone and iPad or Google's Android devices, location based services have been widely used. Currently, the Global Positioning System (GPS) receivers in smartphones primarily provide the location information for these services. Usually, GPS only works well if the smartphone has an unobstructed clear sky view. Besides GPS, smartphones support many additional communications systems such as GSM, UMTS, LTE, WiFi, Bluetooth, and NFC. These communications technologies complement GPS for location based services, especially in urban and indoor environments. Hence, companies like Apple or Google already exploit the identity of WiFi hotspots and cellular base stations for a fast, sometimes crude, first position fix. Besides that, regulatory bodies such as US Federal Communications Commission require operators of cellular communications networks to guarantee, for emergency calls, a service quality of the measured location (FCC 1999). Hence, there exists a strong market and regulation driven development of positioning with current and future wireless communications technologies.

Personally, we started working on localization as early as 1997 with some early signal design studies on Galileo at the German Aerospace Center (DLR). Then, positioning with wireless communications, in particular with cellular communications technologies complementing GPS and the future Galileo system, became our focus from 2005. As our backgrounds are in information and communication theory as well as signal processing, we were not familiar with the specific challenges and requirements of positioning and corresponding signal processing. For example, communications engineers often model the wireless channel by a tapped delay line starting with delay zero. However, they do not take into account the delay of the first arriving path. This path is proportional to the distance between transmitter and receiver. Thus, it is essential to determine the location of a mobile terminal. Hence, this book reflects our learning process on positioning with wireless communications. It also shows our work experience through several projects on positioning.

The content of this book is organized into nine chapters. Through Chapters 1 and 2, the reader will quickly get acquainted with the topic of positioning with wireless communications. Chapter 1 introduces past, current, and future satellite and ground based radio positioning systems as well as critical environments for satellite based positioning systems. Next, Chapter 2 discusses the fundamental positioning principles. These principles are the basis for positioning in today's wireless radio systems.

Then, Chapters 3, 4, and 5 will enable the reader familiar with communication technology or signal processing to achieve a deep technical understanding of the basic positioning technology. Chapter 3 formulates the parameter estimation problem for obtaining position dependent measurements from wireless communications systems. In Chapter 4, positioning algorithms use these measurements to estimate a mobile terminal's location assuming the mobile terminal does not move during the positioning process. Chapter 5 extends the previously static positioning process to dynamic position tracking of moving mobile terminals.

More advanced topics on positioning with wireless communications are addressed in Chapters 6–9. First, Chapter 6 discusses in detail the scenarios and environments in which positioning with satellite and wireless communications systems takes place. It also presents corresponding propagation models for the radio signals and movement models for the mobile user. Second, Chapter 7 presents advanced positioning algorithms such as hybrid data fusion of satellite navigation and positioning with wireless communications, cooperative positioning among mobile terminals, and multipath and non-line-of-sight mitigation concepts. Subsequently, Chapter 8 surveys positioning with various wireless communications systems that are currently widely deployed and in use, or will be in the near future. The book concludes with an introduction to applications of positioning with wireless communications in Chapter 9.

Acknowledgements

The authors would like to thank the many direct and indirect contributors to this book. Many thanks go to Helena Leppäkoski from Tampere University of Technology for allowing us to reproduce her work on WLANs from the Galileo Ready Advanced Mass Market Receiver (GRAMMAR) project. Many thanks also go to Jimmy J. Nielsen from Aalborg University for permitting us to replicate his work on location-aided relay selection and location assisted handover prediction from the Wireless Hybrid Enhanced Mobile Radio Estimators (WHERE) project. Further, we would like to express our sincere thanks to Loïc Brunel, Nicolas Gresset, and Mélanie Plainchault from Mitsubishi Electric R&D Centre Europe, who granted us their permission to reproduce their work on location based inter-cell interference coordination from the WHERE project.

Many thanks to our colleagues from the Mobile Radio Transmission Group, the Department of Communications Systems, and the Institute of Communications and Navigation of DLR for helpful technical discussions. In particular, we thank Simon Plass for helping us on the application of position information for cellular diversity and Wei Wang on positioning with triangulation. Further, we would like to thank the members of the GREAT (Galileo REceiver for mAss markeT), GRAMMAR, WHERE, and WHERE2 project teams, whose work we have cited in this book.

Finally, many thanks to the Wiley team who made this book possible.

List of Abbreviations

2G	Second Generation
3GPP	3rd Generation Partnership Project
3GPP2	3rd Generation Partnership Project 2
A-FLT	Advanced Forward Link Trilateration
AGNSS	Assisted GNSS
AltBOC	Alternative Binary Offset Carrier modulation
AMC	Adaptive Modulation and Coding
AOA	Angle of Arrival
AOAD	AOA Difference
AP	Access Point
API	Application Programming Interface
ASF	Additional Secondary Factors
ASIR-PF	Auxiliary SIR-PF
ATIS	Alliance for Telecommunications Industry Solutions
AWGN	Additive White Gaussian Noise
BCCH	Broadcast Control CHannel
BER	Bit Error Rate
BPSK	Binary Phase Shift Keying
BR	Basic Rate
BS	Base Station
C-CDD	Cellular CDD
C/A	Coarse/Acquisition
CAT	Cellular Alamouti Technique
CBOC	Composite Binary Offset Carrier modulation
CC	Cross-Correlation
CCK	Complementary Code Keying
CDD	Cyclic Delay Diversity
CDF	Cumulative Distribution Function
CDM	Code Division Multiplexing
CEP	Circular Error Probability
CGALIES	Coordination Group on Access to Location Information for Emergency Services
CIR	Channel Impulse Response
CoMP	COordinated Multi-Point

CP	Cooperative Positioning
CPF	Cumulative Probability Function
CPICH	Common Pilot CHannel
CRLB	Cramér–Rao Lower Bound
CRS	Cell-specific Reference Signal
CSI	Channel State Information
CSIT	CSI at the Transmitter
CSMA/CA	Carrier Sense Multiple Access with Collision Avoidance
DAB	Digital Audio Broadcasting
dB	Decibel
DC	Direct Current
DFT	Discrete Fourier Transform
DFTS-OFDM	DFT-Spread OFDM
DGNSS	Differential GNSS
DLL	Delay Locked Loop
DMRS	DeModulation Reference Signal
DOP	Dilution of Precision
DPCCH	Dedicated Physical Control CHannel
DPCH	Dedicated Physical CHannel
DPDCH	Dedicated Physical Data CHannel
DPSK	Differential PSK
DSL	Digital Subscriber Line
DSSS	Direct-Sequence Spread Spectrum
DVB-T	Digital Video Broadcasting–Terrestrial
E-CID	Enhanced Cell-ID
E-IPDL	Enhanced IPDL
E-SMLC	Enhanced Serving Mobile Location Center
E-UTRA	Evolved UTRA
E-UTRAN	Evolved UTRAN
EDAS	EGNOS Data Access Service
EDOP	East DOP
EDR	Enhanced Data Rate
EGC	Equal Gain Combining
EGNOS	European Geostationary Navigation Overlay Service
EKF	Extended KF
EMS	EGNOS Message Server
eNB	Evolved Node B
EOTD	Enhanced Observed Time Difference
EPC	Electronic Product Code
ERP	Extended Rate PHY
ESA	European Space Agency
ETSI	European Telecommunications Standards Institute
FAA	Federal Aviation Administration
FCC	Federal Communications Commission
FDD	Frequency Division Duplex
FDM	Frequency Division Multiplexing

FDOA	Frequency Difference of Arrival
FIM	Fisher Information Matrix
FMC	Fixed Modulation and Coding
GAGAN	GPS Aided Geo Augmented Navigation
GANSS	Galileo and Additional Navigation Satellite Systems
GDOP	Geometric DOP
GEO	Geostationary Orbit
GFSK	Gaussian Frequency-Shift Keying
GLONASS	Global Navigation Satellite System
GLOS	Geometric LOS
GMSK	Gaussian Minimum-Shift Keying
GNSS	Global Navigation Satellite System
GP	Guard Period
GPS	Global Positioning System
GRPR	Golden Received Power Range
GSM	Global System for Mobile Communications
GSO	Geosynchronous Orbit
GTD	Geometric Time Difference
HR/DSSS	High Rate DSSS
HS	HotSpot
HSDPA	High Speed Downlink Packet Access
HSPA+	Evolved High Speed Packet Access
HT	High Throughput
I-WiMAX	3GPP/3GPP2-WiMAX Interworking
I-WLAN	3GPP/3GPP2-WLAN Interworking
i.i.d.	Independent and Identically Distributed
ICIC	Inter-Cell Interference Coordination
IEEE	Institute of Electrical and Electronics Engineers
IFFT	Inverse Fast Fourier Transform
ILS	Instrument Landing System
IP	Internet Protocol
IPDL	Idle Period DownLink
IR	Impulse Radio
IRNSS	Indian Regional Navigational Satellite System
KF	Kalman Filter
LAN	Local Area Network
LBS	Location Based Services
LCS	LoCation Services
LE	Low Energy
LEO	Low Earth Orbit
LFSR	Linear Feedback Shift Register
LMU	Location Measurement Unit
LORAN	LOng Range Aid to Navigation
LOS	Line Of Sight
LPP	LTE Positioning Protocol
LPPa	LTE Positioning Protocol A

LQ	Link Quality
LTE	Long Term Evolution
LTE-A	LTE-Advanced
MAC	Media Access Control
MAI	Multiple Access Interference
MAP	Maximum A Posteriori
MB	MultiBand
MCC	Master Control Center
MCS	Master Control Station
MEO	Medium Earth Orbit
ML	Maximum Likelihood
MMSE	Minimum Mean Square Error
MRC	Maximum Ratio Combining
MSAS	Multifunctional Satellite Augmentation System
MSE	Mean Squared Error
MT	Mobile Terminal
MTCB	MT Clock Bias
NAVSTAR-GPS	Navigational Satellite Timing and Ranging–GPS
NDOP	North DOP
NFC	Near Field Communication
NLES	Navigation Land Earth Station
NLOS	Non-LOS
nm	Nautical Mile
OFDM	Orthogonal Frequency Division Multiplexing
OFDMA	Orthogonal Frequency Division Multiple Access
OTD	Observed Time Difference
OTDOA	Observed TDOA
P-CCPCH	Primary Common Control Physical CHannel
P/Y	Precision/encrYption
P2P	Peer-to-Peer
PBCH	Physical Broadcast CHannel
PDF	Probability Density Function
PDOP	Position DOP
PDP	Power Delay Profile
PF	Particle Filter
PHY	PHYsical layer
PRACH	Physical Random Access CHannel
PRN	Pseudo Random Noise
PRS	Positioning Reference Signal
PSK	Phase-Shift-Keying modulation
PSS	Primary Synchronization Sequence
PUCCH	Physical Uplink Control CHannel
PUSCH	Physical Uplink Shared CHannel
QAM	Quadrature Amplitude Modulation
QPSK	Quadrature Phase-Shift Keying
QZSS	Quasi-Zenith Satellite System

R-PF	Regularized PF
RAIM	Receiver Autonomous Integrity Monitoring
RAN	Radio Access Network
RDC	Reverse Differential Correlation
RF	Radio-Frequency
RFID	Radio-Frequency IDentification
RFPM	RF Patter Matching
RIMS	Ranging and Integrity Monitoring Station
RMS-RX	Root-Mean-Square Received Signal
RMSE	Root Mean Square Error
RRM	Radio Resource Management
RSCP	Received Signal Code Power
RSRP	Reference Signal Received Power
RSRQ	Reference Signal Received Quality
RSS	Received Signal Strength
RSSI	Received Signal Strength Indicator
RSTD	Reference Signal Time Difference
RTD	Real Time Difference
RTTOA	Round-Trip Time Of Arrival
RX	Receiver
SA	Selective Availability
SC-FDMA	Single-Carrier FDMA
SDS-TWR	Symmetric Double Sided Two Way Ranging
SET	SUPL Enabled Terminal
SFN	System Frame Number
SINR	Signal-to-Interference-and-Noise Ratio
SIR-PF	Sampling Importance Resampling PF
SISNeT	Signal-In-Space through the interNET
SLC	SUPL Location Center
SLmAP	SLM Application Protocol
SLP	SUPL Location Platform
SNR	Signal-to-Noise Ratio
SPC	SUPL Positioning Center
SRS	Sounding Reference Signal
SSS	Secondary Synchronization Sequence
STBC	Space-Time Block Code
SUPL	Secure User Plane Localization
TA	Timing Advance
TB	Tail Bit
TDD	Time-Division Duplex
TDMA	Time Division Multiple Access
TDOA	Time Difference of Arrival
TDOP	Time DOP
TIA	Telecommunications Industry Association
TOA	Time Of Arrival
TX	Transmitter

UE	User Equipment
UKF	Unscented KF
UL-RTOA	UpLink Relative Time of Arrival
UMTS	Universal Mobile Telecommunications System
UTC	Coordinated Universal Time
UTDOA	Uplink TDOA
UTOA	Uplink TOA
UTRA	UMTS Terrestrial Radio Access
UTRAN	UMTS Terrestrial Radio Access Network
UWB	Ultra-WideBand
VDOP	Vertical DOP
VLF	Very Low Frequency
W-CDMA	Wideband CDMA
WAAS	Wide Area Augmentation System
WiMAX	Worldwide Interoperability for Microwave Access
WLAN	Wireless LAN
WPAN	Wireless Personal Area Network
WSN	Wireless Sensor Network

1

Introduction

The determination of position is an art that has fascinated scientists for centuries. First positioning methods were probably developed several millennia ago when people realized the necessity of knowing their position for systematic travel. Orientation at natural *landmarks* such as mountains, rivers, or coastlines are straightforward methods for that purpose. Early man made landmarks were trails and ways that were often built for trading, for example, the famous Silk Road, which has its origins around 500 B.C., and connected Europe and Eastern Asia. Other man made landmarks are lighthouses. They provide orientation in monotonous environments even at night, for example, for ships relatively close to the coastline. On the high seas, however, landmarks are missing. Keeping track of a journey by measuring direction and velocity, called the dead reckoning method, was the straightforward approach used by early ocean navigators. Celestial navigation is another method that utilizes well-known objects as position references. Measuring the angle of the pole star above the horizon directly provides the latitude. The major problem for a long time has been the determination of the longitude directly related to the exact measurement of time due to the Earth's rotation. As the Earth rotates around 360° each day, a deviation of 4 s in time keeping results in a position error of 1/60° that is 1 nautical mile or 1.852 km at the equator. At that time, the longitude problem was so severe that several prizes were offered for the development of more precise longitude determination methods. In 1714 the British government rewarded £10 000 for a method capable of determining the longitude within a range of 60 nm (nautical miles), £15 000 for a deviation of 40 nm and £20 000 for 30 nm during a six week journey to the West Indies. Famous scientists like Isaac Newton and Edmond Halley proposed and promoted the use of astronomic methods, that is, predictable astronomic occurrences, for time determination. The 'lunar distance' relative to a fixed star or the ecliptic of Jupiter's moons are such ideas. The invention of chronometers with sufficient accuracy solved the problem and made astronomical methods needless. In 1761 John Harrison's H.4 marine chronometer, constructed in 1759, showed a time deviation of 5 s during a five-week journey to Jamaica. All methods that at least partially rely on visual observations require clear sight. This limits the usability of these methods to certain times of a day or to good weather conditions. The discovery of radio waves in the late nineteenth century opened the door for the field of *radio navigation*. Radio beacons take the role of man made landmarks. Radio frequency bands provide a propagation

Positioning in Wireless Communications Systems, First Edition. Stephan Sand, Armin Dammann and Christian Mensing.
© 2014 John Wiley & Sons, Ltd. Published 2014 by John Wiley & Sons, Ltd.

range exceeding that of visible light. Dependent on the frequency band, radio waves are able to travel through clouds or fog, or even propagate as ground waves over a long distance. This solved the range problem even for ground based radio navigation systems. Nowadays, satellite navigation systems provide global coverage with accuracy in the range of meters. Some of the positioning principles, however, remain the same as for traditional *landmark* or celestial navigation. In particular these are angular methods, where the angle of arrival of radio waves are determined. Today, radio navigation is mainly based on radio propagation time measurements, by which the knowledge of propagation speed (speed of light) provides distance measures related to the radio beacons.

The civil availability of accurate satellite navigation together with chip-sets and navigation receivers for consumer applications have formed the basis of a rapidly growing navigation market in recent years. Indicated by this market growth, the availability of position information will play an increasingly important role in current and future mobile information systems. Information about the position of a user or a *mobile terminal (MT)* can be exploited in a multiplitude of ways. Navigation services for both the consumer and professional market are probably the most well-known applications for positioning systems. Such services can be classified into the following categories:

1. *Positioning*: Determining solely the location of a person or object.
2. *Tracking*: Monitoring the movement of a person or object.
3. *Navigation*: Routing and guidance from an origin to a destination.

These categories are listed regarding increasing usage of auxiliary information and mutual dependency. As an example, tracking requires position determination of a target but usually also incorporates the movement history and a movement prediction of that target in order to achieve a more accurate estimation of the target's trajectory than independent sequential position measurements would. Maps, for instance, provide additional information about environment, especially the traffic infrastructure. This enables route planning, which together with accurate localization and tracking, is the the core of navigation applications. Mobile communications devices are equipped more and more with positioning capabilities that make information about mobiles' positions ubiquitous. The integration of positioning and communications in one device leads to an increasing number of *location based applications and services*. Service providers and end users are not the only ones who can benefit from added value of positioning information. Even network operators can take advantage from the knowledge of mobile devices. Spectrum is an extremely valuable resource and its availability is essential for wireless communications. Information about the position of the mobile communication devices allow an efficient usage of communication resources through the optimization of resource management, handover, or routing procedures.

Algorithms for *communications systems*, which for instance take into account the position of MTs in order to optimize the assignment of radio resources or location and context aware services, are typical examples that show the value of accurate positioning in different layers of a communications system. A simple example, shown in Figure 1.1 points out the added value of position information. An MT moves through an environment covered by a macro-cell with *base station* BS_1 and a pico or femto *hotspot* cell BS_2. The hotspot cell, which could be a *WLAN*, provides a much higher data rate d_2 than the much bigger macrocell does (d_1).

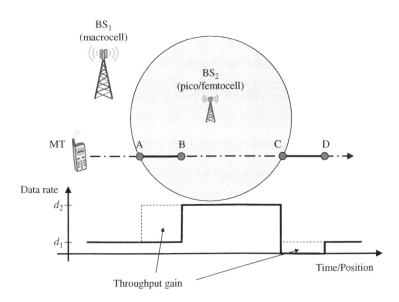

Figure 1.1   Example: *Handover* procedure

At position A of the movement trajectory, the MT becomes aware of the hotspot and starts a handover procedure. This procedure takes some time, during which the MT has moved into the hotspot cell until position B but without getting the higher data rate d_2. This data rate can be exploited from position B until position C, where the MT leaves the hotspot coverage area. Due to the small size of the hotspot, this probably happens too rapidly so there is not enough time for a *seamless handover* back to the macrocell. This handover is completed at position D. The waste of throughput due to handover latency depends on the size of the hotspot cell in relation to the speed of the mobile terminal. Information about the position of both the mobile device and the hotspot area enables us to predict entry into and exit of the hotspot cell and allows us increase the throughput.

Although not primarily designed for positioning, *terrestrial communications systems* can be used to obtain the position of a mobile terminal in a *radio access network (RAN)*. Compared to positioning systems, the main part of transmitted signals are unknown at the receiver. This signal part is information to be transferred from the transmitter to the receiver. Nevertheless, today's *wireless communications systems* specify well-known signal components, called pilots, which are used in a receiver for synchronization and channel estimation purposes. Similar to positioning systems, these signal components can be used for propagation timing measurements and, therefore, positioning. As already mentioned, such systems are designed for communication. Here, the requirements for timing accuracy, in particular synchronization, are usually much weaker than for positioning.

This book focuses on the utilization of terrestrial wireless communications systems for positioning. Before we discuss this as the main topic in the next chapters, we introduce already existing radio positioning systems and environments that are critical for positioning systems to work in.

1.1 Ground Based Positioning Systems

Prior to satellite based positioning systems, radio signals transmitted from terrestrial stations were used for positioning purposes. One challenge for terrestrial positioning systems is to achieve a sufficient coverage and accuracy under certain constraints. Building up dense networks of terrestrial radio beacons either becomes expensive or even impossible, such as on the high seas. For maritime positioning in particular, it is obvious that transmitters for radio positioning have to cover relatively wide areas. Signals radiated in the long wave radio band are well suited to covering large areas. Signals mainly propagate as ground waves, that is, the electromagnetic waves follow the Earth's surface. This allows us to measure their traveling distances, the important figure for ranging, by their traveling times, basically the measurable signal value. This is in contrast to the short wave band, where signals propagate around the globe by sky waves. Sky waves are reflected multiple times between the ionosphere and the Earth's surface, depending on ionospheric properties. However, these reflections are less predictable and make reliable ranging impossible. Therefore, terrestrial wide area radio positioning systems operate at the *long wave band* (30 ... 300 kHz).

1.1.1 DECCA

The DECCA was developed by the British company Decca and deployed during World War II, mainly in the North Sea for maritime navigation in coastal regions. The Allied Forces needed a system for accurate landing operations. It was first used for the landing operation in Normandy in June 1944. On the day prior to D-Day, the first DECCA stations were switched on. Civilian use post World War II has been for fishing vessels or aviation. The system was shut down in Spring 2000.

A DECCA positioning system consists of a number of stations that are organized into so-called chains. A chain consists of a master station and usually three slave stations, which have been termed 'Red', 'Green', and 'Purple'. Geometrically the slave stations are located at the vertices of an equilateral triangle. The master station position is the center of that triangle. The master-slave distance, that is, the baseline length, was about 60--20 nm. Each station transmits a continuous wave signal. A receiver compares the phase difference of the master and a slave signal. Hyperbolas with foci at the master-slave (respectively) positions describe locations of equal signal phase differences. Three hyperbola patterns, associated with the three master-slave pairs 'Red', 'Green', and 'Purple' were drawn on nautical charts. Intersections of the hyperbolas resulting from phase difference measurements provided the position estimate.

It is not desirable for the four stations (one master and three slaves) to transmit their continuous waves using the same frequency. The signals would not have been separable at the receiver in that case. Thus, for simple signal separation, the stations used different frequencies. In order to provide simple phase relations for waveforms the frequencies of the stations in a chain had to be chosen properly. For that, each station used a phase synchronous multiple of a nominal frequency f_0. The nominal frequency was assigned to a chain and, therefore, provided its identification. The harmonics, used by the master and slave stations were $6 \times f_0$ (Master), $5 \times f_0$ (Purple), $8 \times f_0$ (Red), and $9 \times f_0$ (Green). Typical operation frequencies are 70 ... 130 kHz. This *hyperbolic positioning* approach is closely related to the *time difference of arrival (TDOA)* principle, which will be introduced in the next chapter.

The coverage radius of DECCA was around 400 nm (740 km) at daytime, and reduce to 200−−250 nm (460 km) at night. The accuracy of DECCA depended on the angle of intersection of the hyperbolic lines, instrumental errors, propagation errors (e.g., due to sky wave propagation), and the width of lanes. Note, phase differences are measurable in 360° only, which in principal causes ambiguity. A so called lane is the distance of positions corresponding to similar phase differences. In daytime errors ranged from a few meters on the baseline up to a nautical mile at the edge of coverage.

1.1.2   LORAN

The development of the first *LORAN (LOng Range Aid to Navigation)* system was initiated by the US Navy during World War II. It got the name LORAN-A and enabled all-weather navigation for ships and aircraft in the Pacific and North Atlantic Ocean. Its extension, LORAN-B did not go beyond the experimental phase. *LORAN-C* was ready for use in 1957. Since 1958 it has been operated by the US Coast Guard. In 1974 it was decided to phase out the operation of LORAN-A and use LORAN-C as the primary navigation system for the coastal waters of the US and Alaska. LORAN-C stations are located all over the world. LORAN-C covers the North Pacific, the North Atlantic, North Sea, Baltic Sea, Red Sea, Persian Gulf, and so on. The Soviet Union has developed a similar system for its inland called *Chayka*.

Like DECCA, LORAN-C is organized into 'chains', which contain a master station and between two and five slave stations, separated by several hundreds of kilometers. In contrast to DECCA, LORAN-C transmits impulse groups scheduled by a fixed scheme. For the identification of master and slave stations the impulses from different stations are sent out with fixed delays. These emission delays as well as the group repetition interval are chosen so that the signals do not overlap at any location. A LORAN-C signal is transmitted at a carrier frequency of 100 kHz and achieves a coverage of about 1000 km. Figure 1.2 shows a LORAN-C signal. The 100 kHz carrier rapidly increases in amplitude within the first 65 µs as specified in (LOR 1994). Its trailing edge significantly depends on the transmitting sites or equipment. The trailing edge

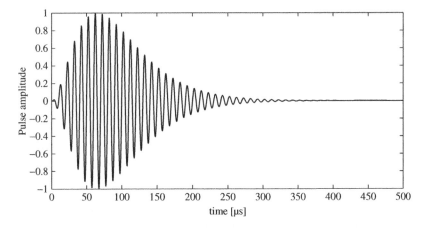

Figure 1.2   LORAN-C pulse

may be controlled in order to meet spectrum requirements. However, after 500 µs the pulse amplitude has to fall below 0.14% (Category 1) and 1.6% (Category 2) respectively, of its peak amplitude. LORAN-C receivers determine propagation time differences. Like DECCA it is a *hyperbolic (TDOA) approach*. The electromagnetic ground wave propagation speed depends on the properties of the Earth's surface. To achieve a good absolute positioning accuracy these *additional secondary factors (ASF)* have to be taken into account. Good receivers achieve an accuracy of some 10 m. Without ASF correction the positioning accuracy is in the range of several hundreds of meters.

1.1.3 OMEGA

The OMEGA radio navigation system was deployed at a time, where a satellite positioning system, TRANSIT, was in operation and plans for GPS have already existed. For that reason it was supposed to be the last ground based positioning system and, therefore, got the last letter of the Greek alphabet. It was used from 1968–1997 for positioning of ships and aircraft with global coverage. Eight OMEGA stations (A, B, ... , H) were distributed around the globe: (A) Aldra, Norway; (B) Monrovia, Liberia; (C) Haiku, Hawaii, USA; (D) La Moure, North Dakota, USA; (E) Réunion, Indian Ocean; (F) Golfo Nuevo, Argentina; (G) Australia; (H) Tsushima, Japan. These stations transmitted signals in the Very Low Frequency (VLF) band (10 ... 14 kHz). Electromagnetic waves in the VLF band penetrate sea water, which meant that the OMEGA system also suitable for submarine navigation. The transmitters achieved a coverage radius of 10 000 nm (18 500 km). Because of that high range, the distance between stations could be up to 5000––6000 nm (9300––1100 km).

The eight OMEGA stations transmitted carriers with frequencies according to a fixed timely scheme. One transmission cycle lasted 10 s. Each station sequentially transmitted four so-called common frequencies (10.2 kHz, 11.05 kHz, 11.33 kHz, 13.6 kHz) and, additionally, a fifth station one that was unique to a single station. OMEGA receivers measured phase differences, thus, this was a *hyperbolic positioning method*. By receiving signals from three stations, an OMEGA receiver could locate a position to within 4 nm (7.4 km).

1.2 Satellite Based Positioning Systems

When talking about positioning systems, people usually have in mind the *Global Positioning System (GPS)* operated by the United States Air Force for the Department of Defense. GPS is probably the most well-known representative of today's *global navigation satellite systems (GNSSs)*. However, there are further satellite based positioning systems in operation, for example, the Russian *Global Navigation Satellite System (GLONASS)* or the Chinese *Beidou* system. Currently planned or built are the European *Galileo* System, *COMPASS* or *Beidou-2* from China, the *Indian Regional Navigational Satellite System (IRNSS)* and the Japanese *Quasi-Zenith Satellite System (QZSS)*. The first space based positioning system was the US *Transit* system. The development of Transit started in 1958. It was in operation from 1964 until 1996.

Predecessors of satellite based positioning systems were ground based systems, for example, *LORAN, DECCA,* or *OMEGA*. Like their space based counterparts, these systems have been designed for positioning purposes and apply related principles for position determination. The very basic principle is the transmission of radio signals that a receiver can use to extract and estimate characteristics related to its position relative to the transmitters. In particular propagation time or time differences of these radio signals are used to extract distances and positions.

The *space segment*, or simply the satellites, is one major component of satellite navigation systems. The choice of *satellite orbits* for space based navigation systems depends on information such as visibility of satellites, coverage, power, launch costs, and so on. *Low earth orbits (LEOs)* with an altitude of less than 2000 km require less launch costs and transmit power. However, a LEO satellite is visible only for some 10 min for each orbit. The number of satellites required for global coverage therefore is high. The influence of the atmosphere on such satellites is high, too, which leads to perturbations of their orbits. On the other hand the *geostationary orbit (GEO)* with an altitude of 36 000 km provides a static satellite visibility but requires higher transmit power and higher launch costs. Since the GEO is above the Earth's equator good coverage can be achieved with a low number of satellites for equator regions. At higher latitudes, however, coverage is poor. *Geosynchronous orbits (GSOs)* like GEO have an orbit time of 24 h. Dependent on their inclination and eccentricity they can provide regional footprints and therefore cover higher altitudes as well. As a good compromise the tree primary satellite navigation systems GPS, GLONASS, and Galileo operate their satellites in *medium earth orbits (MEO)*. The number and distribution of the orbits as well as the number of satellites within each orbit differ between these systems. Figure 1.3 shows the orbits of GPS, GLONASS, and Galileo true to scale. Besides the satellite constellation, further parameters play an important role for the achievable accuracy of GNSSs. Table 1.1 compares basic properties of GPS, GLONASS, and Galileo.

The three satellite systems GPS, GLONASS, and Galileo use the *time of arrival (TOA)* principle for the determination of positions. This principle requires stable and synchronous clocks at the transmitters, which is usually achieved through atomic clocks on board the satellites. From the received signals, the receiver calculates the three coordinates of space and—as a fourth dimension—the navigation system time. With position and time, we have four unknowns, which can be determined using four equations. Behind each equation there is a propagation

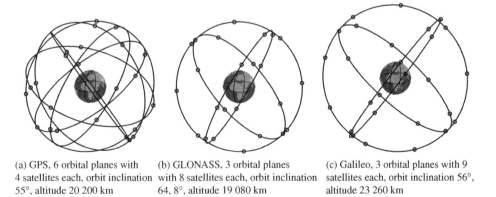

(a) GPS, 6 orbital planes with 4 satellites each, orbit inclination 55°, altitude 20 200 km

(b) GLONASS, 3 orbital planes with 8 satellites each, orbit inclination 64, 8°, altitude 19 080 km

(c) Galileo, 3 orbital planes with 9 satellites each, orbit inclination 56°, altitude 23 260 km

Figure 1.3   GNSS satellite orbits true to scale

Table 1.1   Comparison of global navigation satellite systems

	GPS	GLONASS	Galileo
Number of Satellites	24	24	30
Orbits	6	3	3
Orbit Radius	26 560 km	25 440 km	29 620 km
Orbit Time	11 h 58 min	11 h 15 min 40 s	14 h 5 min
Inclination	55°	64.8°	56°
Multiplex	CDM	FDM	CDM
Code Type	Gold (C/A)	M-sequence (C/A)	Tiered
Code Length	1023 (C/A)	511 (C/A)	4092–10230
Chip Rate	1.023 Mchips/s	0.511 Mchips/s	1.023–10.23 Mchips/s
Nav. Message Length	12 min 30 s	2 min 30 s	10–12 min
Modulation	BPSK	BPSK	BOC, BPSK

time or equivalently a distance measurement to a satellite. Thus, at least four visible satellites are required for the determination of the position.

1.2.1   GPS

The Global Positioning System, the official name is *Navigational Satellite Timing and Ranging – Global Positioning System (NAVSTAR-GPS)*, is a space based positioning system consisting of 24 satellites. It has been developed by the US Department of Defense and, thus, is a military system but can be used for civil applications too. The development of GPS started in 1973. The first GPS satellite has been launched in 1978. In 1993 the 'Initial Operational Capability' was reached with 24 satellites in orbit. The 'Full Operational Capability' was achieved in December 1995. The GPS architecture defines three main entities which are the space segment, the control segment, and the user segment.

1.2.1.1   Space Segment

The space segment consists of the GPS satellites in orbit. GPS originally used three orbital planes with eight satellites in each plane. Later, this was changed to six orbits with four satellites each. The orbits show an inclination of 55° and an equal distribution around the globe resulting in a separation (right ascent) of 60°. Within the orbital planes, the four satellites' positions are unequally distributed. The altitude of 20 200 km results in a orbital time of half a sidereal day. The ground track of a satellite therefore has a period of one (sidereal) day (23 h 56 min). Figure 1.3(a) shows the GPS orbits and satellite positions as a snapshot true to scale.

During the development phase 10 satellites, called Block I satellites, were launched from 1978–1985. The successor generation of these satellites have been the model types Block-II and IIA. The first launch of them was in February 1989. Until 1996, in total nine Block-II and 18 Block-IIA ('A' stands for 'advanced') satellites were brought into orbit. Since then, Block-IIR/IIR-M ('R-M', 'replacement', 'modernized') have successively complemented the system or replaced older satellite types. Block-IIF ('follow-on') are the next generation of satellites. The different satellite generations complemented and modernized the transmitted signals for both civil and military use.

1.2.1.2 Control Segment

The GPS satellites are controlled by the *master control station (MCS)* located at the Schriever Air Force Base near Colorado Springs, Colorado. The MCS's tasks are:

- Monitoring the satellite orbits
- Monitor the satellite state ('satellite health')
- Maintain the system time
- Synchronize the satellite clocks
- Predict and update the satellite orbit models (ephemeris)
- Update the satellite navigation messages
- Perform path correction maneuvers to keep the orbit and relocate satellites.

Monitoring stations are distributed around the globe: Hawaii, Kwajalein, Ascension Island, Diego Garcia, Colorado Springs, and Cape Canaveral. The MCS communicates via S-band (2–4 GHz) ground antennas, which are colocated with most of the monitor stations. MCS, monitor stations, and ground antennas form the Control Segment of the GPS system.

1.2.1.3 User Segment

Equipment that receives and processes the satellites' signals is called the *user segment*. Similar to communications, a GPS receiver first has to capture and transform the *radio frequency (RF)* signal into baseband. Signals from several satellites then have to be separated. For each satellite signal timing and Doppler measurements have to be performed. The timing accuracy directly influences the positioning precision of the receiver. For getting the satellites' orbit data, the navigation messages have to be received, detected, and decoded. Finally, the receiver calculates position, velocity, and system time based on the timing measurements and received navigation data. In the late 1990s navigation receivers were less equipped and expensive. Increasing system integration in electronics led to smaller and cheaper devices with more and more computational capabilities. This enabled the implementation of location based applications on top of the pure position calculation functionalities in one device. Today, GPS chip-sets are available that are integrated in PDAs and cell phones for the mass market.

1.2.1.4 Signals and Carriers

GPS comprises more than 24 satellites, which have to broadcast their navigation signals. Thus, these signals have to share the available radio resources in a proper way so that they can be separated at the receiver side. For that purpose *code division multiplexing (CDM)* using *pseudo random noise (PRN)* codes are used. In GPS two different types of codes, the *C/A (coarse/acquisition)* and the *P/Y (precision/encryption)* code, are specified. The C/A codes are *Gold sequences* (Gold 1967) with a length of 1023 chips. The C/A Gold sequences are constructed by a modulo-2 addition of the binary sequences of two 10-bit *linear feedback shift registers (LFSRs)* with generator polynomials

$$G_1(x) = 1 + x^3 + x^{10},$$

$$G_2(x) = 1 + x^2 + x^3 + x^6 + x^8 + x^9 + x^{10}.$$

The phase shifts between the two component sequences are chosen such that the C/A Gold sequences show both good autocorrelation and cross-correlation properties. Good autocorrelation is necessary for precise timing measurements while cross-correlation determines the orthogonality, that is, the interference of codes transmitted from different satellites. The duration of a C/A sequence is 1 ms which results in a chip rate of 1.023 Mchips/s. Using the speed of light $c \approx 3 \cdot 10^{-8}$ m/s, the length of a C/A sequence is 1 ms $\cdot c \approx 300$ km, which yields a chip length of approximately 300 m. On one hand the C/A code–as its name indicates–is dedicated to the corse acquisition of the P/Y code, which due to its length is much harder to acquire. On the other hand, the C/A code is public usable since its documentation is accessible.

The period of the P/Y-codes is one week. The P/Y-codes are segments of a much longer PRN (approx. 2×10^{14} chips ≈ 266 days) sequence. The chip rate of the P/Y-codes is 10.23 Mchips/s, which is 10 times that of a C/A-code. A P/Y-code consists of the public documented P-code and a Y-code, which is kept secret. The Y part can be switched on or off on demand. As of 1994 the Y-code is permanently switched on.

The position of the satellites is an important parameter that has to be known to the receiver. In GPS these data are transmitted as a binary coded navigation message from the satellites. The data bit-stream modulates the C/A- and P/Y-code sequences by modulo-2 addition in the binary domain. The bit duration of the navigation message is 20 ms that relates to a data rate of 50 bit/s. With this choice, each bit covers 20 C/A-code sequences. It takes about 12.5 min to transmit the whole *navigation message*. Within a navigation message, the important ephemeris and clock parameter data is repeated at a period of 30 s.

The GPS uses several *carrier frequencies (L1-L5) in the L-Band* (1–2 GHz) to transmit the navigation signals that consist of the PRN codes (C/A and P/Y) modulated by the navigation message as described previously. The navigation signals are modulated onto the carriers using *binary phase shift keying (BPSK)*. The following L-band carriers are used respectively planned to be used:

- L1 (1575.42 MHz)   Inphase: P/Y-code
  Quadrature: C/A-code.
- L2 (1227.60 MHz)   Inphase: P/Y-code.
- L3 (1381.05 MHz)   Carries a military signal which supplements the detection of high-energy infrared events such as missile starts or nuclear detonations.
- L4 (1841.40 MHz)   Studied for additional ionospheric correction purposes.
- L5 (1176.45 MHz)   Planned for transmission of a civilian safety-of-life signal.

In spite of being public usable, GPS is primarily a military system. So there is the need to exclude non-authorized users, who are potential military enemies, from a precise positioning service. During the development phase, the performance of the C/A-code has turned out to be surprisingly good. Thus, with the Block-II satellites the performance of C/A-code positioning was artificially degraded. For that, an artificial jitter was added to the C/A-code chip clock. The feature itself is called *Selective Availability (SA)*. With SA turned on a positioning accuracy of about 100 m was achievable. With the switch-off of SA in 2000, the positioning accuracy of the GPS C/A-code increased to 10–15 m.

1.2.2   GLONASS

The GLONASS system is the Russian pendant to the US GPS. During the Cold War, there was the necessity for the Soviet Union, as the leading nation of the Warsaw Pact states, to build up a system similar to GPS that would enable precise navigation for primarily military use. The development of GLONASS started in 1972. Ten years later in 1982, the first GLONASS satellite was launched. The system built up to 26 satellites in orbit until 1995. After that time the number of satellites steadily dropped to seven in 2002. The reasons for that are certainly the end of the Cold War and the collapse of the Soviet Union and the Warsaw Pact. At present time there are 24 satellites in operation. For full operation state, 24 satellites are required, where three of them are spare. GLONASS satellites are arranged in three different MEO planes with an inclination of 64.8° equally distributed around the globe (see Figure 1.3(b)). Within each orbit there are eight satellites equally distributed, that is, the distance of adjacent satellites is 45°. The altitude of the orbits is around 19 100 km, which results in an orbit time of 11 h 15 min. This is 8/17 of a sidereal day. The ground track of the satellites therefore has a period of 17 revolutions or eight (sidereal) days. The satellite constellation provides global coverage. Unlike GPS, GLONASS separates the signals transmitted from different satellites by using different carrier frequencies (GLO 2002). This method is called *frequency division multiplexing (FDM)*. The different carriers at L1 and L2 are

$$\text{L1:} \quad f_{L1} = 1602 + k \times 0.5625 \text{ MHz},$$
$$\text{L2:} \quad f_{L2} = 1246 + k \times 0.4375 \text{ MHz},$$

where $k = -7, -6, \ldots, 13$ is the channel number. Each GLONASS satellite transmits the same sequences. Like in GPS, there is a *C/A-code* and a *P-code*. The C/A code is a *maximum length sequence (M-sequence)* generated from a LFSR with feedback polynomial

$$G(x) = 1 + x^5 + x^9.$$

The C/A code is of length 511 and is transmitted with a chip rate of 511 kchips/s. Thus, the sequence length is 1 ms. The navigation message is transmitted with a data rate of 50 bit/s. Like in GPS, the GLONASS P-code chip rate is 10 times that of the C/A-code.

1.2.3   Galileo

Galileo is a civil satellite navigation system initiated by the European Union and coordinated by the *European Space Agency (ESA)*. The Galileo system comprises 30 satellites in three MEO planes. The orbits are equally distributed around the globe with a right ascent of 120°. Nine satellites in each orbit are separated by 40°. One satellite per orbit is a non-active spare. The altitude of the orbits is 23 260 km. This yields an orbit time of 14 h 5 min. The navigation signals, transmitted from Galileo satellites, are called E1, E6 and E5 with its subsignals E5a and E5b. These signals are located in the L-Band. Figure 1.4 shows their placement in the

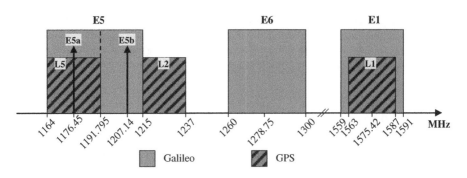

Figure 1.4   GNSS signal spectrum

spectrum in comparison to the L1, L2, and L5 signals of GPS. The full specification of E1 and E5 are public available in (Gal 2010). Like GPS, Galileo uses code division multiplexing for the navigation signals transmitted from the different satellites. These ranging codes are PRN sequences. They can be defined by LFSRs. LFSR based codes are generated as combinations of two M-sequences. Fifty different codes are achieved by using different initial start values for the LFSRs. There are also optimized PRN sequences, which are specified in (Gal 2010). Since these code sequences have to be kept in memory, these codes are often called *memory codes*. These PRN codes represent the *primary codes* of a *tiered code* generation. For this construction method, each period of a primary (binary) code sequence is multiplied by the binary chip value of a *secondary code* sequence. So, the chip length of a secondary code equals the sequence length of the primary code. Secondary codes are memory codes and also specified in (Gal 2010). The ranging codes for the E6 signal are not the subject of this document. Table 1.2 summarizes code lengths and chip rates for the different ranging codes of the freely available Galileo signals.

E1 (1575.42 MHz)
The *E1* signal is composed of two signal components called *E1-B* and *E1-C*. E1-B carries the *I/NAV message* whereas E1-C is a *pilot* component. The modulation is *composite binary offset carrier (CBOC)*. Here, the BPSK modulated E1-B and E1-C chips are further spread by a further binary sequence (subcarrier), with the six-fold subcarrier rate, that is, 6.138 Mchips/s.

Table 1.2   Length and chip rates of the Galileo ranging codes

Signal Component	Tiered Code Length [ms]	Primary Code Chip Rate [Mchips/s]	Code Length [chips]	
			Primary	Secondary
E5a-I	20	10.23	10 320	20
E5a-Q	100	10.23	10 320	100
E5b-I	4	10.23	10 320	4
E5b-Q	100	10.23	10 320	100
E1-B	4	1.023	4092	N/A
E1-C	100	1.023	4092	25

The E1 signal supports *Safety-of-Live*, *system integrity*, and the *open service*. The receiver reference bandwidth of the E1 signal is 24.552 MHz.

E5a (1176.45 MHz)

The *E5a* signal is one part of the *E5* signal and overlaps in the spectrum with the GPS L5 signal. It consists of a data part E5a-I, which is transmitted in the inphase component of E5a and carries the *F/NAV message*. The quadrature component E5a-Q carries a *pilot* sequence without any data modulation. The receiver reference bandwidth of the E5a signal is 20.46 MHz.

E5b (1207.14 MHz)

The *E5b* signal is a second part of the E5 signal. It consists of a data part E5b-I, which is transmitted in the inphase component of E5a and carries the *I/NAV message*. The quadrature component E5b-Q carries a *pilot* sequence without any data modulation. The receiver reference bandwidth of the E5b signal is 20.46 MHz.

E5

The signal *E5* consists of the two components E5a and E5b. The modulation is *alternative BOC (AltBOC)*, which alternatively can be described as *8-PSK (phase-shift-keying)* modulation with restrictions in possible state transitions. The receiver reference bandwidth of the E5 signal is 51.15 MHz.

E6 (1278.75 MHz)

The *E6* signal is composed of two signal components called E6-B and E6-C. As for E1, these signal parts are modulated onto the inphase component. In contrast to E1, the modulation is *BPSK*. The E6-B part carries the *C/NAV message* whereas E6-C is a *pilot*. The receiver reference bandwidth of the E6 signal is 40.92 MHz.

The Galileo signals introduced here carry different navigation messages. There are three different types of these data formats, supporting different services:

F/NAV

This navigation message supports the freely accessible open service. Its carrier is the E5a-I signal component. One message frame has a length of 600 s and carries the satellites' ephemeris and almanac, as well as clock correction and ionospheric correction data.

I/NAV

This message carries system integrity information and supports the Safety-of-Life service. The carrier of this message is the E5b-I and E1-B signal. One message frame has a length of 720 s.

C/NAV

This is the commercial navigation message. The carrier of this message is the E6-B. The C/NAV data format is not part of the Galileo signal in the space interface control document (Gal 2010).

1.3   GNSS Augmentation Systems

Navigation satellites are transmitting signals, that is, electromagnetic waveforms, whose structures are known to the receiver. In principle, this allows us to measure the traveling distance of these signals from the satellite to the navigation receiver. These propagation time measurements are the basis for the calculation of the receiver position. There are several sources of error. These errors can be basically categorized into two types:

- *Randomly occurring errors*: These kind of errors show a low level of correlation or statistical dependencies. Thermal noise for instance is an example for such errors. Averaging over several measurements is one strategy to combat these kind of errors.
- *Systematic errors*: These errors typically show high correlation in several dimensions, for example, in location and time.

The correlation properties of systematic errors can be exploited to combat or at least reduce them. Clock errors in navigation satellites are obviously systematic. They are equal for all navigation receivers for a certain period of time, that is, there is a strong local and temporal correlation. The *ionosphere* is another source of systematic errors. The ionosphere is the uppermost part of the atmosphere beginning at a height of approximately 80 km and finally fading to interplanetary space. It contains free electrons and electrically charged atoms and molecules. This ionization is primarily caused by ultraviolet solar radiation. Free electrons and ions severely influence the electromagnetic propagation speed, that is, the speed of light. However, this parameter is required for calculation of the distance of a navigation receiver to a satellite. Not taking into account the ionospheric signal delays causes positioning errors in the range of several hundreds of meters. The ionospheric delay is usually combated by appropriate prediction models or the usage of multi-frequency receivers. However, residual errors remain. It is obvious that these errors show local correlation since the electromagnetic waves from a satellite to receivers in a certain area travel through the same part of the ionosphere.

Systematic errors can be measured by a reference station and reported to navigation receivers. This is the principle of augmentation systems. In the following we will briefly introduce such augmentation systems, distinguishing between local systems like *differential GNSS (DGNSS)*, which is a representative of *ground based augmentation systems (GBASs)* or *space based augmentation systems (SBASs)* like the European EGNOS or the US WAAS system. These systems supply much wider areas.

1.3.1 Differential GNSS–DGNSS

Positioning accuracy requirements of different applications vary widely. Some tens of meters of accuracy are sufficient for navigation in wide and open areas, like maritime navigation at the high seas or en route aircraft guidance. For such applications GNSS positioning itself is appropriate. For accuracy requirements in the range of meters or even below, the DGNSS approach allows us to further reduce systematic system errors.

Some parts of positioning errors are similar for navigation receivers that are close enough to each other. Additionally they do not vary too much in time. This is what we have also previously termed local and temporal correlation. These spatio-temporal correlations are exploited at a DGNSS receiver by subtracting an estimated systematic error value provided by a DGNSS reference station.

Figure 1.5 illustrates the DGNSS principle. Having a mean value of zero, randomly occurring errors, for instance caused by thermal noise, can be reduced by averaging. Since a DGNSS reference station is fixed, averaging can be done for a long period of time, allowing us to

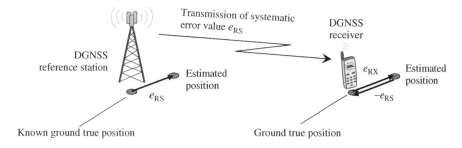

Figure 1.5   Principle of DGNSS

significantly reduce these kind of errors. For simplicity, we assume these errors to be zero in Figure 1.5.

We assume that the ground true position of the DGNSS reference station is known exactly. A GNSS receiver at this reference station additionally estimates the position of that station. Most likely, there will be a deviation of the estimated position compared to the ground true position. This difference is mainly caused by systematic errors, which are the same at GNSS receivers in the neighborhood of the DGNSS reference station. Subtracting the ground true position from that estimated at the DGNSS reference station yields the systematic error vector e_{RS} at the reference station. This error vector is transmitted via appropriate data links to DGNSS capable receivers surrounding the reference station. Due to spatial and temporal correlations $e_{RX} \approx e_{RS}$. The remaining systematic error after subtraction of e_{RS} from the initial position estimation at the DGNSS receiver, $e = e_{RX} - e_{RS} \approx 0$, is significantly reduced. Because of building differences, errors in the estimation of e_{RS} directly propagate into the final position estimation accuracy. Thus, random error terms in e_{RS} should be reduced as much as possible at the reference station.

1.3.2   Wide Area Augmentation System – WAAS

The US Wide Area Augmentation System is a satellite based augmentation system for improving accuracy, integrity, and availability of GPS. Starting in 1994, WAAS was developed by the United States Department of Transportation and the *Federal Aviation Administration (FAA)*. The initial goal was to provide accuracies comparable to a Category I *Instrument Landing System (ILS)*. After switch-off of the C/A selective availability, which artificially decreased positioning accuracy, the *ionospheric signal delay* effects remained the main source of error. WAAS provides correction data, which allows us to significantly reduce these errors. Specifications requirements state a positioning accuracy of 7.6 m for 95% of the time. Measurements show an accuracy around 1 m laterally and 1.5 m vertically for most parts of the USA, Canada, and Alaska. A further task of the system is to detect errors in the GPS/WAAS network. These are, for instance, erroneous signals that are misleading and, therefore, bear potential safety risks. WAAS aims to warn users about integrity problems within 6.2 s.

The system consists of a network of reference stations in North America and Hawaii. Measurements from these ground stations are processed in order to get deviation correction values. Via master stations these values are routed to geostationary satellites, which broadcast them back to Earth. WAAS signals are transmitted in the L1-band using GPS-like modulation, that

Table 1.3   SBAS satellites

SBAS System	Satellite	Satellite Position	PRN Code #
WAAS	Inmarsat 4-F3	98° W	133
	Intelsat Galaxy XV	133° W	135
	TeleSat Anik F1R	107.3° W	138
EGNOS	Inmarsat 3-F2	15.5° W	120
	Artemis	21.5° E	124
	Inmarsat 4-F2	25° E	126
	SES-5	5° E	136
MSAS	MTSAT-1R	140° E	129
	MTSAT-2	145° E	137
GAGAN	GSAT-8	55° E	127
	GSAT-10	83° E	128

is, a BPSK modulated C/A sequence with a data rate of 250 bits/s. Half-rate convolutional coding results in a symbol (encoded bits) rate of 500 bits/s. This allows to receive these signals without additional overhead in the RF part of receivers. WAAS enabled GPS receivers use these correction data for improving the accuracy of position estimation following the differential GNSS principle. Coverage area of WAAS is the North American continent. Table 1.3 lists the geostationary satellites transmitting the WAAS signal together with further SBAS systems that will be briefly introduced in the following.

1.3.3   European Geostationary Navigation Overlay Service – EGNOS

EGNOS is a European satellite based augmentation system for the US GPS, the Russian GLONASS, and the European Galileo systems. It is a joint project of the European Union and ESA. EGNOS distributes correction data via geostationary satellites and is compatible with the US WAAS and the Japanese MSAS systems. Reference stations, *ranging and integrity monitoring stations (RIMSs)*, receive the positioning signals from GPS, GLONASS, and (later on) Galileo satellites. These RIMSs are distributed over the whole of Europe and North Africa and allow us to obtain comprehensive information about the ionosphere all over Europe. From the RIMSs' data, four *Master Control Centers (MCCs)* calculate correction data for the satellite orbits as well as maps containing information about the electron density in the atmosphere. Uplink stations, *navigation land earth stations (NLESs)*, transmit these information to geostationary telecommunications satellites (see Table 1.3), which broadcast them free of charge. Like WAAS, EGNOS transmits its data using the L1-band and GPS C/A codes with identification numbers beyond 32[1].

Transmitted data sets are archived and provided via the *EGNOS message server (EMS)* (EMS n.d.) by FTP. This allows post processing for GNSS raw data recordings. The much larger amount of RIMS raw data is provided by the *EGNOS data access service (EDAS)* (EDA n.d.), which is a commercial service. EGNOS, like WAAS, is mostly designed for aviation users

---

[1] Regular GPS satellites use C/A codes with IDs 1 … 32

who are able to receive signals from geostationary satellites directly and undistorted. On the ground, however, the usage of EGNOS is limited due to the relative low elevation of geostationary satellites. To solve this problem, ESA released an Internet based service called *SISNeT (signal-in-space through the Internet)* (SIS 2002) in 2002. This service continuously delivers EGNOS data to ground users.

An initial operation phase of EGNOS started in July 2005. Certification for safety-of-life applications was achieved in 2011 (ESA 2011). EGNOS specifies an accuracy of better than 7 m. In practise, accuracy is around 1 m with a system availability of greater than 99%.

1.3.4   Multi-Functional Satellite Augmentation System–MSAS

MSAS is the Japanese pendant to the European EGNOS and US WAAS systems. Two geostationary satellites, MTSAT (see Table 1.3) support meteorological and communications services. Supporting GPS at the moment, these satellites broadcast correction and integrity data for Japan. Carrier frequency and modulation are similar to WAAS and EGNOS. The signal is transmitted in the L1-band using BPSK modulation with a data rate (information bits) of 250 bits/s, which yields an convolutionally encoded bit rate of 500 bits/s. C/A codes with IDs 129 and 137 are assigned to MSAS. With an accuracy around 1–2 m in both horizontal and vertical directions, the performance of MSAS is comparable to WAAS and EGNOS. The first MSAS satellite, MTSAT-1R, was successfully launched in February 2005, followed by MTSAT-2 one year later. MSAS has been operational since September 2007.

1.3.5   GPS Aided Geo Augmented Navigation–GAGAN

Supporting GPS the GAGAN system provides satellite based correction signals for India. Its intention is to provide navigation service which meets the requirements for all phases in flight over Indian airspace. The system comprises eight reference stations located in Delhi, Guwahati, Kolkata, Ahmedabad, Thiruvananthapuram, Bangalore, Jammu, and Port Blair. A master control center is located at Bangalore. The Indian government intends to create its own regional navigation system, the *IRNSS*. For India, GAGAN is the first technological step towards an autonomous satellite navigation system. In 2007 the initial 'Technology Demonstration System' phase was completed. The first satellite carrying the GAGAN payload, GSAT-8, was launched in May 2011. GSAT-10, the second GAGAN satellite, followed in September 2012.

1.4   Critical Environments

In the first step we need to define what is meant by critical environments. Using radio signals for positioning we face the problem how to relate received waveforms to the position of the receiver. At the receiver we only have the opportunity to measure physical parameters of those waveforms. A relation of received signals to the receiver position becomes feasible when these signal parameters can be predicted to some extent. To make more clear what we mean by that, let us have a look at the *TOA principle* as an example. For this positioning principle, we need to know the distance between a transmitter, for example, a satellite, and the receiver. However, this is a parameter that we cannot measure directly. Instead of the transmitter-receiver distance we are measuring the propagation time of a signal traveling from the transmitter to the receiver. To relate this signal propagation time to the required distance, we have to make some assumptions. Essential assumptions for that are:

- Electromagnetic waves travel on the shortest path from the transmitter to the receiver, that is, we have *line of sight (LOS)* propagation conditions. It is obvious that reflected or diffracted electromagnetic waves reaching the receiver violate that assumption.
- The traveling speed of electromagnetic waves from the transmitter to the receiver is known. The *speed of light* in vacuum is $c \approx 3 \times 10^8$ m/s. However, the signal propagation speed, for example, in the ionosphere, may significantly differ from c.

The effects mentioned here are strongly dependent on the environment, which makes them challenging to predict or estimate. Another straightforward problem is the strength of the received signal. Especially indoors, GNSS signals are weak. For this reason, it is hard to achieve sufficient accuracy. To summarize, we face a critical environment whenever we cannot adequately predict or model measurable signal parameters to locations. Typical examples are:

- *Urban canyons*, which typically provide *non-LOS (NLOS)* and *multipath propagation*.
- *Indoor environments*, which suffer from NLOS propagation and sever signal attenuation. NLOS occurs when electromagnetic waves penetrate buildings. These signals often face a severe attenuation, which makes them useless for indoor positioning.

In urban areas, buildings, especially in narrow streets, restrict the direct view to sky. This restricts the number of satellites that can be received with LOS conditions. Figure 1.6 illustrates an urban scenario. It is obvious that the height of the buildings in relation to the street width strongly influences the LOS visibility of the satellites. Some signals are received through some detour. This NLOS propagation can be caused by reflection or diffraction at the buildings.

Figure 1.7 provides a visibility snapshot for GPS satellites in Munich downtown, obtained by a ray tracing simulation. The results show less than four visible satellites for large street sections. Note at least four satellites are necessary for position estimation.

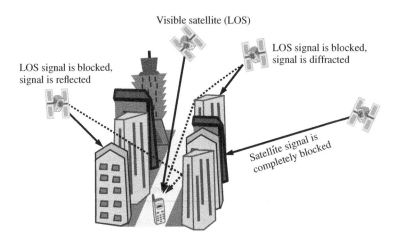

Figure 1.6 Urban canyon environment: Signal propagation paths may become longer due to reflection or diffraction

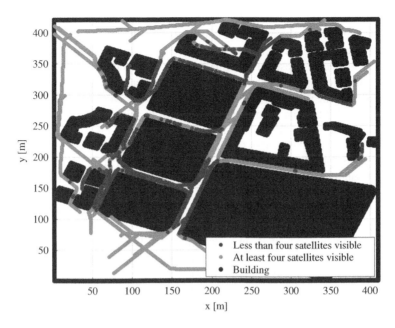

Figure 1.7   Visibility of GPS satellites in an urban environment (Munich downtown)

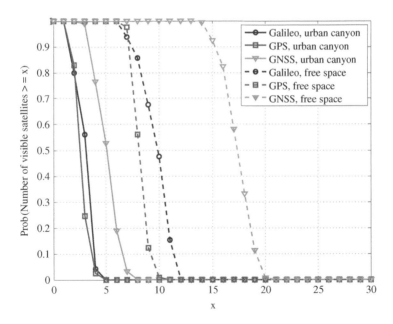

Figure 1.8   Cumulative probability function for GPS and Galileo satellite visibility in an urban canyon and open field

Figure 1.8 shows the *cumulative probability function (CPF)* for GPS and Galileo LOS visibility in two different environments. The results have been obtained by ray tracing simulations. Dashed lines represent an open field scenario. There are no obstacles, which limit the view to the sky. This yields the best case for satellite visibility. With a probability of one, at least four satellites are visible for both GPS and Galileo. It can be seen that Galileo outperforms GPS in terms of visibility in the range of $\geq 7 \ldots 12$ visible satellites. With a probability of 0.56, ≥ 8 GPS satellites are visible in an open field scenario. For the Galileo satellite constellation, this probability is 0.86. A hybrid GPS and Galileo receiver would observe ≥ 15 satellites with a probability of 0.92. The hybrid GPS+Galileo system is termed GNSS in that plot.

The situation completely changes in an urban canyon environment. The probability of observing ≥ 4 satellites is 0.02 and 0.03 for GPS and Galileo. At least four satellites are necessary for calculating a position. This means that the failure probability in an urban canyon is 0.98 respectively 0.97. Even if both systems are combined, the failure probability, that is, less than three satellites are visible, is 0.23.

These results show that, especially in critical scenarios like the urban canyon environment, GNSS positioning requires complementary positioning systems. A promising approach is the use of terrestrial wireless communications systems. In urban areas communications infrastructure is especially dense, which makes them suitable for positioning purposes. In the following chapters of this book, we will introduce algorithms and methods for positioning with wireless communications systems.

2

Positioning Principles

Positioning in general requires an observation of physical quantities. These measurement values characterize a certain position in the environment. Thus, it is evident that observations for positioning purposes have to show local dependency. In this chapter we focus on positioning principles, which are typically applied when observing electromagnetic waveforms for localization purposes. Geometric parameters significantly influence the propagation of electromagnetic signals:

1. *Propagation delay*: Electromagnetic waves travel with the speed of light. This yields a linear relation between the signal propagation delay and the distance of transmitter and receiver.
2. *Propagation attenuation*: The electromagnetic signal power decreases as the distance between transmitter and receiver increases. In free space the received power is reciprocally proportional to the square of that distance.
3. *Reflection, refraction, diffraction, scattering*: Discontinuities in the propagation medium cause changes in the propagation direction of electromagnetic waves. Such discontinuities are omnipresent in typical radio communications environments and cause *NLOS* and *multipath propagation*.

We may state that there is a functional relationship

$$f : \quad (s, x, t) \mapsto r,$$

which maps a waveform s transmitted at the origin of a Cartesian coordinate system at time t to an observation r, received at position $x = (x, y, z)$. The observation r is typically corrupted by additive noise, which is usually modeled as a Gaussian random process. From the noisy observation r the unknown position x of the mobile unit has to be estimated.

The description of the function f can be more or less complex, dependent on the specific assumption about the propagation environment, that is, the propagation model we are going to make. In NLOS propagation environments, knowledge of the position and properties of reflectors, scatterers, and so on is necessary to describe a rather complex function f. In this

Positioning in Wireless Communications Systems, First Edition. Stephan Sand, Armin Dammann and Christian Mensing.
© 2014 John Wiley & Sons, Ltd. Published 2014 by John Wiley & Sons, Ltd.

case the function f is provided by a lookup table (database). Such databases can be set up by reference measurements or signal propagation simulations, for example, ray tracing, and they contain characteristic values. These values are the fingerprints of their associated positions and a receiver tries to find the fingerprint respectively the position of best match. Therefore, such positioning methods are called *fingerprinting* methods.

The function f becomes simple in the case of LOS propagation in free space. However, it is not invertible, since lots of positions, that is, those lying on a circle or sphere around the transmitter show equal propagation delay and attenuation. Taking into account observations from different transmitter locations provides unambiguous positioning. It should be mentioned that a observed propagation time or received signal power is certainly a fingerprint of a specific location as well. The methods for calculating the position are based on the functional dependency of observation and position given the assumed propagation model rather than on the pattern recognition strategies for database methods mentioned previously.

Radio location methods require a *transmitter (TX)* and *receiver (RX)*. There are two approaches for position determination:

1. *Self-positioning*: A signal is transmitted from a TX station at a known position and received by a RX unit whose position has to be determined. The position is calculated at the device to localize itself.
2. *Network-positioning*: A signal is transmitted by a TX unit whose position has to be determined and received from an RX station at a known position. The position is calculated at a unit in the network of RX stations.

Broadcasting systems like *DVB-T* or *DAB* are unidirectional and, therefore, can be used for self-positioning. Mobile communications systems like *GSM (Global System for Mobile Communications)*, *UMTS (Universal Mobile Telecommunications System)* or *3GPP-LTE (3rd Generation Partnership Project's Long Term Evolution)* are bidirectional, that is, electromagnetic waveforms are transmitted from a BS to an MT (downlink) and vice versa (uplink). Therefore, both self-positioning and network-positioning are applicable.

2.1 Propagation Time

In the following the transmitters are BSs of a wireless communications system. The receiver, whose position is to be estimated, is an MT within this system. So we are exemplarily considering the downlink of a mobile communications system. As mentioned, the principles can be applied reversely in the uplink.

The subsequently introduced positioning methods are based upon the measurement of geometric distances between the BSs and a MT. These distances are measured indirectly by the corresponding signal propagation delays. Therefore two basic assumptions have to be made:

1. The signal propagation speed has to be known.
2. Electromagnetic waves propagate along the shortest path, that is, there is no reflection, refraction, diffraction, and so on, that cause a signal detour.

2.1.1 Time of Arrival – TOA

2.1.1.1 Principle

The *TOA* uses measurements of the signal propagation delay in order to calculate the distance of a receiver to a transmitter. The signal propagation distance from BS_i

$$d_i = \int_P ds = \int_{T_i}^{T_0} c\, dt = c_{av}(T_0 - T_i) \overset{\substack{c\ \text{is}\\ \text{constant}}}{=} c(T_0 - T_i)$$

to the MT is calculated by integration along the propagation path P from the speed of light c along that path, the time of transmission T_0, and the time of reception T_i. Knowledge of the propagation speed profile along the propagation path is hard to obtain. Thus, we assume that we know an equivalent average speed c_{av}. In this case the signal propagation distance depends on the propagation delay $T_i - T_0$. It is easy to see that $c_{av} = c$ if the signal propagation speed c along the propagation path is constant.

Assuming LOS propagation, the signal propagation distance d_i determines points of equal distance of the MT to BS_i. If we consider the two-dimensional case, this is the definition of a circle with radius d_i around BS_i. In order to obtain a unique position, the distances to several BSs have to be measured. The intersection of the corresponding circles provides a unique position of the MT. Figure 2.1 illustrates the TOA principle for the two-dimensional case. The intersection of two circles yields two possible solutions. A distance measurement to a third BS resolves that ambiguity.

The distance measurements to N BSs form a system of N *nonlinear equations*. The unknown values of this system are the components of the MT position (x, y, z).

$$\sqrt{(x - x_1)^2 + (y - y_1)^2 + (z - z_1)^2} = c(T_1 - T_0) = d_1,$$

$$\sqrt{(x - x_2)^2 + (y - y_2)^2 + (z - z_2)^2} = c(T_2 - T_0) = d_2,$$

$$\vdots \tag{2.1}$$

$$\sqrt{(x - x_N)^2 + (y - y_N)^2 + (z - z_N)^2} = c(T_N - T_0) = d_N.$$

Here, (x_i, y_i, z_i) denotes the position of BS_i. $T_i - T_0$ is the signal propagation delay corresponding to the signal propagation distance d_i. The equation system (2.1) considers three dimensions. For the two-dimensional case the equation system is easy to adapt by simply skipping the terms $(z - z_i)^2$. For the two (x, y) or three (x, y, z) unknowns of the MT position we would require two or three equations. However, due to nonlinear properties, one more equation is required as mentioned earlier, in order to resolve ambiguous solutions.

In practical systems, the measurement of the distances d_i are noisy in general. Therefore a solution of the equation system (2.1) does not exist in general. This is shown in Figure 2.1. Here, the measurement of the propagation delay is corrupted by an error ϵ. Geometrically, this provides the dashed circle. It can be seen that there is no point where all three circles intersect.

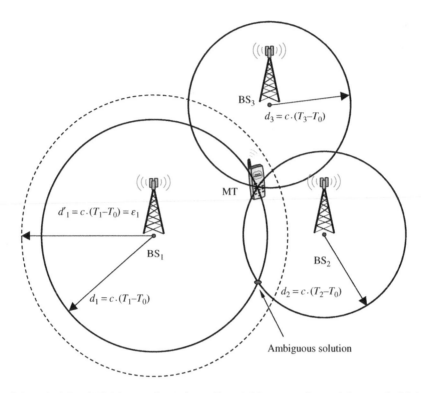

Figure 2.1 Principle of TOA in two dimensions: Given LOS propagation and the speed of light c, the measurement of propagation delays T_i provide the distances (circles) to the BSs. The intersection of these circles yields the MT position

The chance of not finding a solution increases even more if the equation system is overdetermined, that is, there are more equations than unknowns. A usual approach is to consider the equivalent system of equations

$$(x - x_i)^2 + (y - y_i)^2 + (z - z_i)^2 - c^2(T_i - T_0)^2 = \epsilon_i^2, \qquad i = 1, \dots, N \qquad (2.2)$$

that introduces an error value ϵ_i per equation. An estimation

$$(\hat{x}, \hat{y}, \hat{z}) = \arg \min_{(x,y,z)} \sum_{i=1}^{N} \epsilon_i^2 \qquad (2.3)$$

of the MT position is obtained by minimizing the total quadratic error.

2.1.1.2 Problems of TOA

The signal propagation distances d_i are calculated from the corresponding signal propagation delays $(T_1 - T_0)$. In order to obtain the signal propagation delay, the time bases in all the

BSs and the MT must be the same. This requirement is hard to achieve. On the network side, some effort can be spent synchronizing the time bases of the BSs and applying highly stable clocks. At least the asynchronism between the BSs can be measured and provided in order to compensate that deviation. So quasi synchronism among the BSs can be achieved with reasonable effort. However, the problem of different time bases between the network side (BSs) and the MT remains since mobile devices, at least mass market devices, utilize low cost hardware components. Such components provide reasonable short-term stability only. These oscillators are stable enough during the measurement of the times of reception $T_1, \ldots, T_N$. Therefore, the time base T_0 of the BSs is a further unknown value at the MT. Alternatively, the *clock offset* or *bias* $b = c(T_0 - T_M)$ can be used as unknown value at the MT. Using an additional equation allows us to resolve this variable.

Previously, we have mentioned that the measurement of the signal propagation delay is corrupted. In practical systems, thermal noise is omnipresent. It is usually modeled as an additive Gaussian distributed random process. Its mean value is zero. Therefore, this kind of error is called unbiased. Figure 2.2 illustrates unbiased additive noise as a tube around the mean value $c(T_i - T_0)$. The standard deviation of the noise term directly influences the accuracy of the position solution.

NLOS propagation caused by reflection, refraction, or diffraction is another significant source of error. In terrestrial wireless communications, NLOS propagation is a typical phenomenon. The electromagnetic waves do not propagate from the transmitter to the receiver on the shortest possible path. Therefore, we measure a propagation delay that is larger than that of the LOS path. Figure 2.3 shows signal reflection as a source of errors. If the MT is not aware of the reflector, it assumes LOS propagation. The signal propagation delay corresponds to the total signal travel distance $d'_1 = d_{11} + d_{12} = c(T_1 - T_0) \geq d_1$. In case the position of the reflector is known, the distance between the BS and the reflector d_{11} can be calculated and subtracted from the total signal travel distance. This difference $d_{12} = c(T_1 - T_0) - d_{11}$ is the radius of a circle around the reflector and provides an unbiased position estimation.

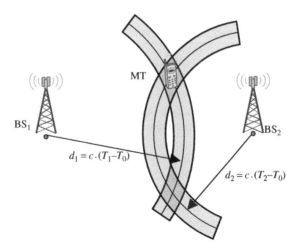

Figure 2.2   Unbiased error as an additive random process term with zero mean

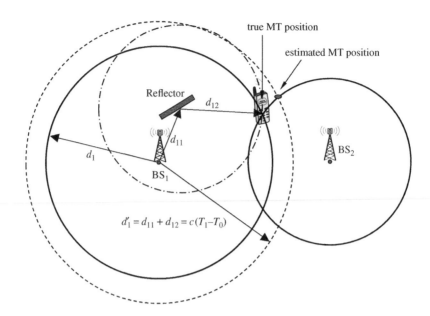

Figure 2.3   Biased error caused by reflection

2.1.2   Time Difference of Arrival–TDOA

The *TDOA* principle is another signal propagation delay based method. The idea for TDOA is to measure signal propagation delay differences. Let us assume that two signals are transmitted from two BSs i and j synchronously at time instance T_0. At the receiver we receive the signal from BSs i and j at time instances T_i and T_j respectively. The corresponding propagation distance difference

$$\Delta d_{i,j} = d_i - d_j = c(T_i - T_0) - c(T_j - T_0) = c(T_i - T_j) = c \cdot \Delta T_{i,j}$$

is calculated from the propagation delays. It is obvious that the propagation distance difference depends on the signal propagation delay difference $\Delta T_{i,j}$. Due to building differences, we get rid of the unknown time T_0 of signal transmission and we do not face problems due to different time scales at the BSs and the MT. Both time instances T_i and T_j that are used for calculating the signal propagation delay difference are measured at the MT and thus result from the same time base.

   In contrast to TOA, where a propagation delay defines a circle around a BS, a signal propagation delay difference measurement of TDOA characterizes points of equal distance differences to the considered BSs. This is the definition of a hyperbola in the two-dimensional case or a hyperboloid for the three-dimensional space. Hence, TDOA is often called hyperbolic positioning. Figure 2.4 shows the principle of TDOA positioning. Hyperbolas with foci at the positions of the involved BSs define geometric points of equal distance difference to those BSs. The intersection of different hyperbolas, or hyperboloids in the three-dimensional case, provide the MT position. In the example, shown in Figure 2.4, the intersection of two hyperbolas (solid lines) provide a unique MT position. A further hyperbola (dashed line) is defined

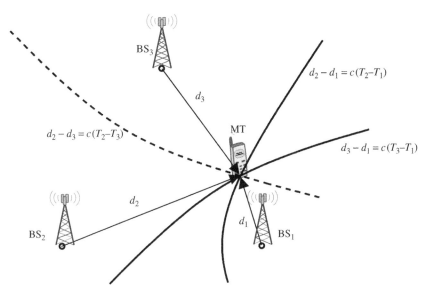

Figure 2.4 Principle of TDOA in two dimensions: Given LOS propagation and the speed of light c, the measurement of propagation delay differences $T_i - T_j$ provide hyperbolas with BSs as foci. The intersection of these hyperbolas yields the MT position

by $d_2 - d_3 = (d_2 - d_1) - (d_3 - d_1) = c\Delta T_{2,1} - c\Delta T_{3,1}$. However, this hyperbola is obtained by signal propagation delay difference measurements from the first two.

The system of $N - 1$ nonlinear equations is then

$$\sqrt{(x - x_2)^2 + (y - y_2)^2 + (z - z_2)^2}$$
$$- \sqrt{(x - x_1)^2 + (y - y_1)^2 + (z - z_1)^2} = d_2 - d_1 = c\Delta T_{2,1},$$

$$\sqrt{(x - x_3)^2 + (y - y_3)^2 + (z - z_3)^2}$$
$$- \sqrt{(x - x_1)^2 + (y - y_1)^2 + (z - z_1)^2} = d_3 - d_1 = c\Delta T_{3,1}, \qquad (2.4)$$

$$\vdots$$

$$\sqrt{(x - x_N)^2 + (y - y_N)^2 + (z - z_N)^2}$$
$$- \sqrt{(x - x_1)^2 + (y - y_1)^2 + (z - z_1)^2} = d_N - d_1 = c\Delta T_{N,1}.$$

The position of the MT and BS_i is denoted by (x, y, z) and (x_i, y_i, z_i) respectively. For the two-dimensional case, the terms containing $(z - z_i)$ are skipped. Obviously, we can obtain the equation system (2.4) from equation system (2.1) by subtracting one equation from the remaining ones. In this example we have chosen the first one for subtraction. Compared to TOA this step resolves the unknown BS time base T_0. With N BSs, we obtain a set of $N - 1$ independent equations.

As for TOA, the measurements $\Delta T_{i,j}$ are noisy. Furthermore, there may be more equations than unknowns. The equation system can be solved using a least squares algorithm similar to Equations (2.2) and (2.3) for the TOA principle.

For the TDOA approach propagation distance differences are calculated from time measurements. Therefore the necessary assumptions are similar to those of TOA. On the one hand the signal propagation speed has to be known. On the other hand we have to assume LOS propagation. If there is NLOS propagation, the position of reflectors or obstacles causing refraction or diffraction is required. Otherwise these positions are further unknowns in the system of TDOA equations.

Example 2.1.1 *To clarify the hyperbolic structure of positions with equal delay differences, we consider the two-dimensional case and choose the coordinate system such that two BSs–say $i = 2$ and $j = 1$–are located symmetrically on the x-axis at positions $(\pm a, 0)$. From the first equation of the Equation System (2.4) we get*

$$\sqrt{(x-a)^2 + y^2} - \sqrt{(x+a)^2 + y^2} = c\Delta T_{2,1}.$$

By squaring this equation twice we arrive at

$$\frac{x^2}{\left(\frac{c\Delta T_{2,1}}{2}\right)^2} - \frac{y^2}{a^2 - \left(\frac{c\Delta T_{2,1}}{2}\right)^2} = 1. \tag{2.5}$$

Based on the measurement of the TDOA $T_2 - T_1$, Equation (2.5) describes hyperbolas with foci in the locations $(\pm a, 0)$ of the involved BSs.

2.1.3   Round-Trip Time of Arrival–RTTOA

Like TOA, the *round-trip time of arrival (RTTOA)* is a circular positioning method. RTTOA measures the signal round-trip delay between two entities and, therefore, requires bidirectional communications capabilities. Figure 2.5 illustrates the signal flow for the RTTOA principle.

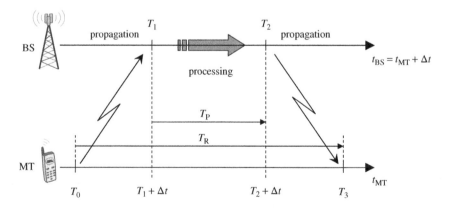

Figure 2.5   Principle of RTTOA in two dimensions: Given LOS propagation and the speed of light c, the measurement of round-trip propagation delay $c((T_3 - T_2) + (T_1 - T_0))$ provides the distance between the two entities

In this example the involved communications entities are a BS and an MT. Here, an RTTOA measurement is initiated by the MT at time instance T_0, measured according to the MT's time scale t_{MT}. The signal is received at the second terminal, that is, the BS, at time instance T_1. The time scales t_{MT} and t_{BS} of the MT and the BS are assumed to be different with constant difference $t_{BS} - t_{MT} = \Delta t$. The signal propagation distance between the MT and the BS then becomes

$$d_{1,2} = c((T_1 + \Delta t) - T_0). \tag{2.6}$$

Here, $T_1 + \Delta t$ denotes time instance T_1 relative to the time scale of the MT. A further signal is sent back from the BS to the MT beginning at time index T_2 relative to the time scale of the BS. The signal is received at the MT at time T_3. The propagation distance is

$$d_{2,1} = c(T_3 - (T_2 + \Delta t)). \tag{2.7}$$

We assume that the geometry between the two signal transmissions does not change and a channel is reciprocal. This means $d_{1,2} = d_{2,1}$. Summing up Equations (2.6) and (2.7) yields

$$d = d_{1,2} = d_{2,1} = \frac{d_{1,2} + d_{2,1}}{2} = \frac{1}{2}c(T_R - T_P). \tag{2.8}$$

Equation (2.8) depends on two time differences. The round-trip time $T_R = (T_3 - T_0)$ is calculated at the MT. $T_P = (T_2 - T_1)$ is the signal processing time, measured at the BS. Both time differences are determined with respect to the respective time bases. Therefore, the signal propagation distance according to Equation (2.8) does not depend on an unknown difference Δt between both time references.

Previously, we started to discuss the problem of different time bases by assuming a constant difference between them. In practical systems the *clocks* of the RTTOA entities may run at different speeds. This means that they show a drift in time. With *linear drifts* α_{MT} and α_{BS} for the MT and the BS, the time bases of the two entities related to a exact reference t are

$$t_{MT} = (1 + \alpha_{MT})t + \Delta t_{MT},$$
$$t_{BS} = (1 + \alpha_{BS})t + \Delta t_{BS}. \tag{2.9}$$

Using the drift model in Equation (2.9) we can rewrite Equation (2.8) and get

$$d_{MBM} = \frac{1}{2}c((1 + \alpha_{MT})T_R - (1 + \alpha_{BS})T_P)$$
$$= \frac{1}{2}c(T_R - T_P)\underbrace{\frac{((1 + \alpha_{MT})T_R - (1 + \alpha_{BS})T_P)}{T_R - T_P}}_{\text{error term}}. \tag{2.10}$$

Note that T_R and T_P are now given with respect to the reference time base t. We observe a multiplicative error term, which is dependent on the drifts α_{MT} and α_{BS} weighted by the round-trip time T_R and the processing time T_P.

The RTTOA measurement was initiated by the MT. Another RTTOA measurement d_{BMB} may be initiated by the BS. Similar to Equation (2.10) we get

$$d_{BMB} = \frac{1}{2}c((1 + \alpha_{BS})T_R - (1 + \alpha_{MT})T_P). \tag{2.11}$$

Averaging both Equations (2.10) and (2.11) yields a distance

$$d = \frac{d_{\text{MBM}} + d_{\text{BMB}}}{2} = \frac{1}{2}c(T_R - T_P)\underbrace{\left(1 + \frac{\alpha_{\text{MT}} + \alpha_{\text{BS}}}{2}\right)}_{\text{error term}} \qquad (2.12)$$

with an error term that does not depend on T_R and T_P. This method is called *symmetric double sided two way ranging (SDS-TWR)* (Nan 2007). The error term in Equation (2.12) contains an average of the drifts α_{MT} and α_{BS}. In case the drifts are equal in their absolute value but with opposite signs, the influence of the error term vanishes.

We may face the situation where one of the time bases is by far more precise compared to the other one. For our example it is a reasonable assumption that the clock at the BS is more stable than that of the MT. Therefore, we get

$$d = d_{\text{MBM}} = \frac{1}{2}c\left(\frac{T'_R}{(1 + \alpha_{\text{MT}})} - T_P\right),$$

$$d = d_{\text{BMB}} = \frac{1}{2}c\left(T_R - \frac{T'_P}{(1 + \alpha_{\text{MT}})}\right), \qquad (2.13)$$

where T'_R and T'_P are the less precise measurements of the round-trip time and the processing time, respectively, made at the MT. We correct them by multiplication with the inverse of the drift factor $(1 + \alpha_{\text{MT}})$. We can eliminate the MT drift in Equation (2.13) by calculating

$$d_{\text{MBM}}T'_P + d_{\text{BMB}}T'_R = d(T'_R + T'_P). \qquad (2.14)$$

From Equations (2.13) and (2.14) we finally obtain

$$d = \frac{1}{2}c\left(\frac{T_R T'_R - T_P T'_P}{T'_R + T'_P}\right)$$

for the distance between MT and BS.

2.1.4 Comparison of Circular and Hyperbolic Positioning

In the previous sections we have introduced positioning principles that are based on signal propagation delay measurements. We have seen that using a propagation delay measurement itself, that is, the TOA principle, provides circles around a reference point, that is, a BS. In contrast to that, signal propagation delay differences, used by TDOA, provide hyperbolas with the BSs as foci. The common approach for both is that we have to calculate the intersection of the respective curves (or surfaces in case of three dimensions). Those nonlinear equation systems show different properties. We compare the circular (TOA, RTTOA) and hyperbolic (TDOA) positioning principles graphically in Figure 2.6. For both circular and hyperbolic positioning, we have drawn circles respectively to hyperbolas, which divide the baseline between two BSs into 20 intervals of equal size. This illustrates an uncertainty of 5% of the BS distance.

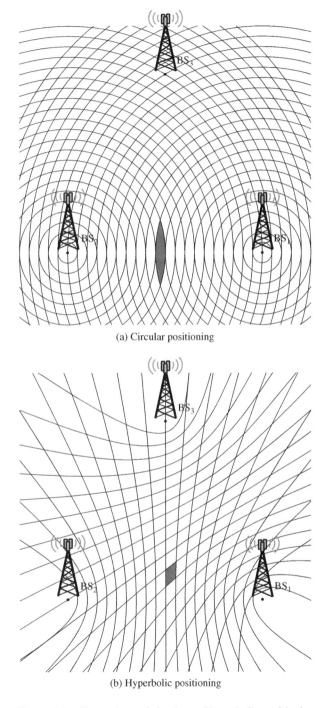

(a) Circular positioning

(b) Hyperbolic positioning

Figure 2.6   Comparison of circular and hyperbolic positioning

For circular positioning, as shown in Figure 2.6(a), regions of uncertainty show quite different shapes, for example, an elongate form between two BSs compared to a more quadratic form in the core of the region spanned by $BS_1 - BS_3$. Particularly in the middle of the baseline between BSs, the uncertainty in the direction orthogonal to this baseline is quite high. In contrast, the hyperbolic positioning principle provides rhombic shapes with more balanced uncertainty in different directions. In particular, this is the case in the middle of a baseline. An uncertainty area for both circular and hyperbolic positioning is highlighted in gray and shows the difference in the proximity to a baseline.

Hyperbolic positioning, as shown in Figure 2.6(b), requires three BSs. Signal propagation delay differences in our example are calculated for the BS pairs $BS_1 - BS_2$ and $BS_1 - BS_3$. For circular positioning, we have to intersect two circles, which requires two BSs. A third BS is required in order to resolve the unknown *clock bias* between the the network of BSs and the MT. Another significant difference between circular and hyperbolic positioning is the ambiguity of position solutions. Circles around BSs intersect at two different points. Both points are symmetric to the baseline between the respective BSs. In the proximity of a baseline these two position solutions are close to each other. We may require a further propagation delay measurement, that is, a further BS, in order to resolve that ambiguity. For the hyperbolic positioning example, we observe that the hyperbolas intersect in a unique point. Therefore, they do not bear the problem of ambiguous position solutions.

2.2   Angle of Arrival–AOA

2.2.1   Two-Dimensional

Previously, we have discussed positioning principles, which are based on the measurements of distances to reference stations (BSs). Another parameter, which is dependent on the position, is the direction from which the signal arrives at the receiver. This results in the *angle of arrival (AOA)* positioning principle. The AOA method can be applied in both downlink and uplink. Subsequently, we will consider the more general case of downlink AOA. Figure 2.7 illustrates the AOA positioning method for the two-dimensional case in the downlink. Here, AOA values are determined at the MT. Given LOS propagation, the measurement of the direction φ_i of the received signals and the BS position (x_i, y_i) determines a straight line, describing possible locations of the MT. For our purpose, we may use a polar coordinate description. For a BS_i, we get

$$\begin{pmatrix} x - x_i \\ y - y_i \end{pmatrix} = r_i \begin{pmatrix} \cos(\varphi_i) \\ \sin(\varphi_i) \end{pmatrix}, \tag{2.15}$$

which can be rewritten as

$$y - y_i = \tan(\varphi_i)(x - x_i)$$

by dividing the components of Equation (2.15). Thus, we get also rid of the distance r_i between the MT and BS_i. The intersection of two lines provides the MT position. Hence, we have to solve the linear equation system

$$\begin{aligned} y - y_1 &= \tan(\varphi_1)(x - x_1) \\ y - y_2 &= \tan(\varphi_2)(x - x_2) \end{aligned} \tag{2.16}$$

for the position (x, y) of the MT.

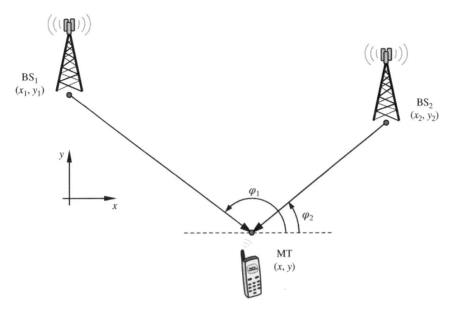

Figure 2.7   Principle of AOA in two dimensions: Given LOS propagation the measurement of the direction of the received signal (angle of arrival) determines a straight line, describing possible locations of the MT

Previously, we have measured the AOAs with respect to the reference coordinate system, in which the positions of the BSs are described. Therefore, we have implicitly assumed that the orientation of the MT with respect to this coordinate system is known at the MT. In Figure 2.7 this orientation, which is parallel to the x-axis, is drawn as a dashed line. Figure 2.8 illustrates the problem of an unknown receiver orientation. The AOA difference $\Delta\varphi = \varphi_1 - \varphi_2$ is independent of the unknown MT orientation φ. According to the inscribed angle theorem, the angle $\Delta\varphi$ and the two BS positions determine the major arc of a circle. For an unknown orientation φ the MT position may be anywhere on that arc. An AOA measurement to a further BS solves that ambiguity.

Including the unknown orientation φ into Equations (2.15)–(2.16) yields the set of equations

$$y - y_i = \tan(\varphi_i + \varphi)(x - x_i), \qquad i = 1, \ldots, N,$$

which is nonlinear in the unknown orientation φ.

2.2.2   Three-Dimensional

To obtain a three-dimensional position solution, we may determine both azimuth φ_i and elevation ϑ_i of a signal arriving from BS_i. Figure 2.9 shows the AOA principle for three dimensions. To keep that example simple, we assume for the moment that the orientation of the MT is known. The measurement of the azimuth φ_1 of the signal arriving from BS_1 determines a plane. Azimuth φ_2 and elevation ϑ_2 of the signal arriving from BS_2 determines a straight line. The intersection of both geometries yields the MT position in three-dimensional space.

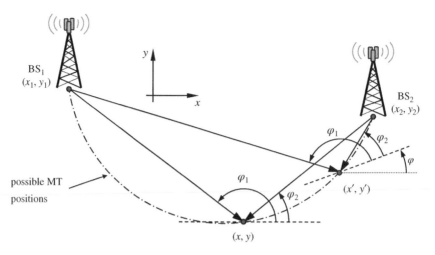

Figure 2.8   Influence of the MT orientation φ: Different values of φ yield different position solutions

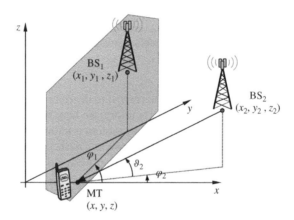

Figure 2.9   Principle of AOA in three dimensions: The measurement of the azimuth φ_1 of the signal arriving from BS_1 determines a plane. Azimuth φ_2 and elevation ϑ_2 of the signal arriving from BS_2 determines a straight line. The intersection of both geometries yields the MT position in three-dimensional space

Using spherical coordinates we obtain

$$\begin{pmatrix} x - x_i \\ y - y_i \\ z - z_i \end{pmatrix} = r_i \begin{pmatrix} \cos(\vartheta_i + \vartheta)\cos(\varphi_i + \varphi) \\ \cos(\vartheta_i + \vartheta)\sin(\varphi_i + \varphi) \\ \sin(\vartheta_i + \vartheta) \end{pmatrix},$$

which we can rewrite to, for example,

$$(y - y_i) = \tan(\varphi_i + \varphi)(x - x_i)$$
$$\cos(\varphi_i + \varphi)(z - z_i) = \tan(\vartheta_i + \vartheta)(x - x_i),$$

(2.17)

in order to eliminate the radius r. For each BS we get a set of two equations according to Equation (2.17). The three components of the MT position (x, y, z) as well as the MT orientation (φ, ϑ) are unknown. Thus, we need at least five equations, which can be obtained from the measurement of azimuth and elevation to at least three BSs.

2.2.3   AOA in the Uplink

Previously, we have discussed the AOA principle applied in the downlink. We have seen that the orientation of the MT is critical. In case the orientation is not known, it must be estimated. For the application of AOA in the uplink, that is, the AOA values are determined at the BSs, this problem is much simpler. In general, the BSs are fixed in their position. It can be reasonably assumed that the orientations of the receivers, that is, the BSs are known. The equations derived previously still hold with the special case of known orientations φ and ϑ.

We have not yet addressed the methodology of the AOA measurement itself. However, this requires a directional antenna, which can be realized by an antenna array. Due to the limited size of an MT, the number of antenna array elements used for downlink AOA is restricted. At the BS, the size of the antenna and, therefore, the achievable directional resolution is a minor problem.

2.2.4   The Problem of Non-Line-of-Sight Propagation

The AOA principle exploits the propagation direction of received signals in order to determine a position. Phenomenons like reflection, refraction, and diffraction significantly influence the propagation direction of electromagnetic waves. Such effects can cause significant errors in the position solution, especially when obstacles like reflectors are located in the near of the receiver. Figure 2.10 illustrates the problem of NLOS propagation for the application of AOA in the uplink. The problem in the downlink is similar. Here, a reflector acts as a mirror and produces a virtual image of the MT. The AOA for the signal, transmitted from the MT and measured at BS_1, results in a totally wrong position estimate.

2.3   Fingerprinting

In this section, we are going to discuss the idea of fingerprinting positioning. As a fingerprint is unique to a human, the properties of a signal are typical for the position where it is transmitted or received. So, we can state that a fingerprint in wireless positioning is the set of measurable signal characteristics that depend on the position of transmission or reception. In the previous sections of this chapter, we have already exploited characteristics such as the signal propagation time (TOA, RTTOA), the propagation time difference (TDOA), or the propagation direction (AOA). Those positioning principles can therefore be considered as fingerprinting methods as well. They assign signal characteristics like propagation time to a position by a relation, which can be analytically described. For instance, we map a signal propagation delay to a circle with a specific radius around the BS.

However, there are further signal characteristics where we do not have a simple analytical relationship because of a complex signal propagation environment. The general approach in this case is to build up a *database*, which contains the fingerprint, that is, the set of signal

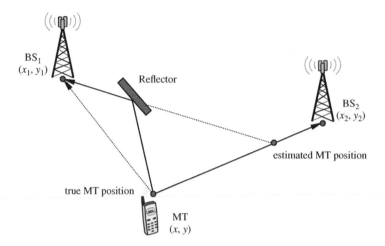

Figure 2.10   The NLOS problem of AOA: A reflector changes the AOA value, which results in erroneous positioning estimation

characteristics, for each position in the environment. A fingerprinting method then compares a measured fingerprint to the fingerprint entries of that database in order to determine the database fingerprint that matches best. This best match then provides the most likely position. Such databases can be built up by measurements or by using simulative methods like ray tracing. Subsequently, we are going to introduce fingerprinting methods based on the *received signal strength (RSS)* or the *power delay profile (PDP)*.

2.3.1   Cell-ID

One of the most simplest and straightforward positioning methods is the *Cell-ID* principle. The basic idea is to decide whether a BS's signal is present or not. Dependent on that decision, the Cell-ID principle concludes that the MT is located within the coverage area of a BS or outside. The size and shape of this area (cell) can either be estimated or determined more precisely by measurements. The only information, which has to be obtained from the signal, is the ID of the device that transmitted the signal. Figure 2.11 shows an example for Cell-ID positioning. The MT receives signals from three BSs. The conclusion is that its position must be within the coverage areas of all three BSs at the same time. Therefore, the possible position solution is somewhere within the intersection area

$$A_{\text{MT}} = \bigcap_{i=1}^{N_{\text{BS}}} A_{\text{BS}_i}$$

of the coverage areas A_{BS_i} of the N_{BS} received BSs' signals. The positioning accuracy of the Cell-ID principle depends on

- the size of the coverage areas of the BSs,
- the correctness (reliability) of their determination, and

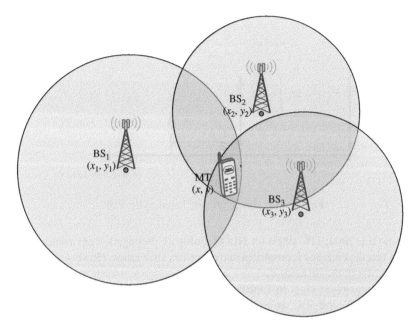

Figure 2.11 The Cell-ID principle: Receiving the signal from a BS results in the conclusion that the MT is located within the coverage area of the BS

- the number of received BSs because

$$\bigcap_{i=1}^{N_{BS}} A_{BS_i} \leq \bigcap_{i=1}^{N_{BS}-1} A_{BS_i}.$$

With the Cell-ID principle, we measure a received signal power and perform a binary classification as to whether the MT is located within the coverage area of a BS or not. A generalization of that principle will be introduced subsequently.

2.3.2 Received Signal Strength–RSS

Previously, we have seen that the Cell-ID classifies the signal power of the received signal binary. Allowing a quantization higher than that of Cell-ID yields the *RSS* principle. The achievable granularity of the quantization is limited by the applied receiver hardware, in particular the analogue-to-digital converter. In general it can be observed that the average received signal power is inversely dependent on the distance d between transmitter and receiver. Commonly used channel models (IEE 2004; JTC 1993) describe that dependency as path loss. The proportionality of the received power is

$$P_{RX} \propto P_{TX}\, G_{TX}\, G_{RX} \left(\frac{d}{d_0}\right)^{-\beta}.$$

β is called the decay factor and depends strongly on the environment. Its value is at least 2 in the case of free space propagation. Values up to 3.5–3.8 can be found in channel model

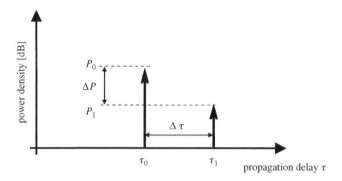

Figure 2.12   Fingerprinting based on the PDP

specifications (IEE 2004; JTC 1993) for NLOS multipath propagation environments. It should be noted that this dependency is empirical and describes an average. However, it allows at least a rough estimation of the distance d.

Factors like P_{TX}, G_{TX}, G_{RX} in Equation (2.17), that is, the transmitted power and the antenna gains of TX and RX, as well as a proportionality constant linearly influence the received signal power. These values have to be known in order to use absolute measures of the received signal power according to Equation (2.17). The problem of an unknown proportionality constant is also present for database-supported RSS positioning. Here, we face the problem that a RSS database is usually built up by some measurement equipment and used later on for positioning by lots of different mobile devices. The scaling of the RSS database values must be the same for both the measurement device, which builds up the database, and the mobiles that use that database for positioning. This requires a calibration in order to achieve a common and unique scaling.

2.3.3   Power Delay Profile–PDP

We can further generalize the fingerprinting positioning principles by using several characteristics of the received signal. One of these characteristics is the received signal power, which we have discussed in the previous section on the RSS positioning method. Multipath propagation causes reception of several replicas of the transmitted signal. These replicas arrive at different propagation delays and with different signal strengths. In Section 2.3.2, we have mentioned that positioning based on absolute power measurements may be difficult to achieve because of unknown proportionality constants or the lack of receiver calibration. We have a similar problem for the measurement of the absolute propagation delays because this would require synchronism of the time bases at the transmitter and receiver. However, we can obtain relative values, which means time differences and power ratios between arriving paths. Figure 2.12 shows a PDP and illustrates a propagation delay difference measurement $\Delta\tau = \tau_0 - \tau_1$ and a propagation path power ratio measurement $\Delta P = P_1 - P_0$. Note, for the power we use a logarithmic notation (dB), which converts the power ratio into a difference. With the approach of relative measurements we get rid of normalization and synchronization problems.

Figure 2.13 shows an example, which illustrates PDP fingerprinting positioning. We consider a rectangular room (Figure 2.13a), where a BS is located at the lower left corner at position

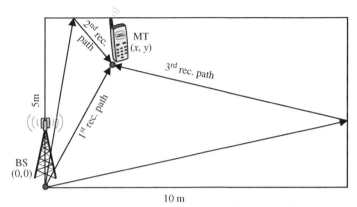

(a) Example scenario: Propagation paths with one reflection at most

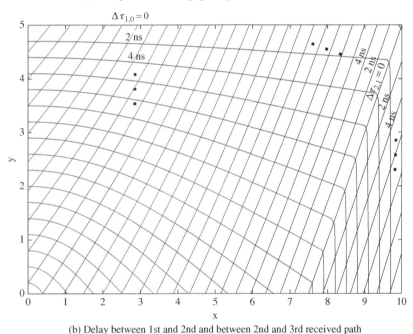

(b) Delay between 1st and 2nd and between 2nd and 3rd received path

Figure 2.13   Fingerprinting example based on the PDP, in particular, the temporal distribution of the received paths

$(0, 0)$. We use a ray tracing approach to calculate the signal propagation delays at the MT position (x, y) within that room.

Figure 2.13(b) shows the locations for propagation delay difference measurement $\Delta\tau_{1,0} = \tau_1 - \tau_0$ between the first and second arriving paths in steps of 2 ns. Similar the propagation delay difference measurement $\Delta\tau_{2,1} = \tau_2 - \tau_1$ between the second and third arriving paths are plotted. For this simple analysis, we assume that we can resolve all the three paths, regardless of their delay difference. At specific positions, these differences approach zero, which means that we cannot resolve those paths in practical systems. This is the case for $\Delta\tau_{1,0}$ at positions $x \to 10\,\text{m}$ and $y \to 5\,\text{m}$. For $\Delta\tau_{2,1}$ we face this situation at $x \to y$. Especially when having a look on $\Delta\tau_{2,1}$ it is obvious that this causes ambiguous position solutions. Close to $\Delta\tau_{2,1}$ we observe contour lines with the same value. For instance, there are two lines for $\Delta\tau_{2,1} = 2\,\text{ns}$ close to each other, causing ambiguity.

For this example, we have required one BS only, which is an advantage of fingerprinting positioning solutions compared to methods that use propagation delay or directional approaches. In practice, however, the situation is much more complex than the one we have exemplarily shown previously. Usually the propagation environment contains lots of reflectors or scatterers, which makes the calculation or prediction of the PDP too complex. Therefore, the use of databases that are built up by measurement data is a common solution to provide a relation between position and received signal characteristics.

In this example, we have used two propagation delay difference measurements for fingerprinting. It is obvious that including more characteristic values into the fingerprints improves both positioning accuracy and the ambiguity problem. For positioning accuracy it is desirable that the fingerprint characteristic values change rapidly with the position. A more rapid change causes higher 'distances' between fingerprints of adjacent positions. Higher distance will result in a higher robustness against measurement noise.

3

Measurements and Parameter Extraction

Wireless positioning is based on the observation of radio signals. Several parameters of a received radio signal depend on the position of the receiver. So, a first step in wireless positioning is the estimation of those signal parameters that depend on location. Such parameters for instance are delay, amplitude, or phase of a signal. These parameters have to be derived from an observed signal. Normally, this signal is represented by noisy signal samples. Here, noise is a kind of signal distortion, which we do not know deterministically. However, we can consider such distortions as stochastic processes where information about the parameters of the stochastic process may be known or can be estimated. A well-known example for stochastic signal distortions is additive white Gaussian noise (AWGN).

3.1 Parameter Estimation

We are interested in obtaining a parameter, which has a continuous range of values, from a received set of signal samples. Subsequently, we will formulate the estimation problem in general. The quality of parameter estimation is of great interest. Parameters are estimated from a received signal, which is corrupted stochastically. Thus, estimates of the parameters itself will be noisy as well. We will consider the variance of estimates as a quality measure. It is obvious to raise the question about the optimum achievable estimation performance. The Cramér–Rao inequality provides a lower bound on the variance of an estimate, meaning that no unbiased estimation algorithm can perform better than this bound.

3.1.1 The Estimation Problem

We observe or measure a set of k values, assembled in a vector $r = [r_1, \ldots, r_k]$. These values depend on a set of ℓ parameters $\alpha = [\alpha_1, \ldots, \alpha_\ell]$. The problem of parameter estimation is to find a vector $\hat{\alpha}(r) = [\hat{\alpha}_1(r), \ldots, \hat{\alpha}_\ell(r)]$, the estimate of the parameters α that depend on the observations r. The observations are noisy and, therefore, random variables. For this reason,

Positioning in Wireless Communications Systems, First Edition. Stephan Sand, Armin Dammann and Christian Mensing.
© 2014 John Wiley & Sons, Ltd. Published 2014 by John Wiley & Sons, Ltd.

the estimates $\hat{\alpha}(r)$ that depend on them, are random variables too. We describe the dependency between the parameters α and the random observations r in terms of the conditional probability density function (PDF) p($r|\alpha$).

An estimate $\hat{\alpha}(r)$ is called unbiased if its (conditional) expectation

$$E_{r|\alpha}\left\{\hat{\alpha}(r)\right\} = \int \hat{\alpha}(r)\,p(r|\alpha)\,dr = \alpha \tag{3.1}$$

equals the parameter α, which is to be estimated. The estimate is said to be consistent if it converges to α as the number of observations k increases. Mathematically, consistency means, that for every small number ϵ, the estimate $\hat{\alpha}(r) = \hat{\alpha}([r_1, \dots, r_k]) = \hat{\alpha}_k$ remains within a distance ϵ to parameter α with probability one as k gets sufficiently large, that is,

$$\lim_{k \to \infty} P\left[\left|\hat{\alpha}_k - \alpha\right| < \epsilon\right] = 1, \qquad \forall \epsilon > 0. \tag{3.2}$$

Example 3.1.1 (Estimation of mean and variance) *As an example let us consider k independent and identically distributed (i.i.d.) real valued random variables $r_i \in \mathbb{R}$, $i = 1, \dots, k$ with mean E $\{r_i\} = \mu$ and variance E $\{(r_i - \mu)^2\} = \sigma^2$, which are obtained according to their conditional PDF p $\left(r_i|\mu, \sigma^2\right)$. An example for such a PDF is the Gaussian distribution*

$$p\left(r_i|\mu, \sigma^2\right) = \frac{1}{\sqrt{2\pi\sigma^2}} e^{-\frac{(r_i-\mu)^2}{2\sigma^2}}, \qquad i = 1, \dots, k.$$

From the k observations, r_i, we would like to obtain an estimate $\hat{\mu}_k$ of the mean value μ. For this, we consider the sample average

$$\hat{\mu}_k = \frac{1}{k} \sum_{i=1}^{k} r_i. \tag{3.3}$$

Its mean value is

$$E\left\{\hat{\mu}_k\right\} = E\left\{\frac{1}{k}\sum_{i=1}^{k} r_i\right\} = \frac{1}{k}\sum_{i=1}^{k} E\left\{r_i\right\} = \frac{1}{k}\sum_{i=1}^{k} \mu = \mu.$$

We have shown that the estimate defined in Equation (3.3) fulfils Equation (3.1) and, therefore, is unbiased. Let us now have a look whether consistency holds for that estimator. First, we calculate the variance of the estimate $\hat{\mu}_k$

$$\sigma_{\hat{\mu}_k}^2 = E\left\{(\hat{\mu}_k - \mu)^2\right\} = E\left\{\frac{1}{k^2}\sum_{i=1}^{k}\sum_{j=1}^{k}(r_i - \mu)(r_j - \mu)\right\}$$

$$= \frac{1}{k^2}\sum_{i=1}^{k} E\left\{(r_i - \mu)^2\right\} = \frac{\sigma^2}{k}, \tag{3.4}$$

where we have used E $\{(r_i - \mu)(r_j - \mu)\} = E\{(r_i - \mu)\}E\{(r_j - \mu)\} = 0$ for $i \neq j$. This means that the variance of the estimate $\hat{\mu}_k$ decreases as k increases. In order to lower bound the probability of $\hat{\mu}_k$ remaining in an interval around μ we use Chebyshev's inequality (Proakis 2001)

$$1 \geq P\left[\left|\hat{\mu}_k - \mu\right| < \epsilon\right] \geq 1 - \frac{\sigma_{\hat{\mu}_k}^2}{\epsilon^2} = 1 - \frac{\sigma^2}{k^2\epsilon^2}, \tag{3.5}$$

where the left side inequality is the trivial one. It is easy to see that $\lim_{k\to\infty} 1 - \frac{\sigma^2}{k^2\epsilon^2} = 1$ and, therefore, Equation (3.2) holds.

In order to demonstrate a biased estimator, we consider an estimator for the variance of the random process described earlier,

$$\hat{\sigma}_k^2 = \frac{1}{k}\sum_{i=1}^{k}(r_i - \hat{\mu}_k)^2, \tag{3.6}$$

which uses an estimate of the mean value according to Equation (3.3). We calculate the expectation

$$\mathrm{E}\left\{\hat{\sigma}_k^2\right\} = \frac{1}{k}\sum_{i=1}^{k}\mathrm{E}\left\{\left(r_i - \frac{1}{k}\sum_{n=1}^{k}r_n\right)^2\right\} = \frac{1}{k^3}\sum_{i=1}^{k}\sum_{n=1}^{k}\sum_{m=1}^{k}\mathrm{E}\left\{(r_i - r_n)(r_i - r_m)\right\}$$

$$= \frac{k-1}{k^2}\sum_{i=1}^{k}\left(\mathrm{E}\left\{r_i^2\right\} - \mu^2\right) = \frac{k-1}{k}\sigma^2,$$

where we have used $\mathrm{E}\left\{r_i^2\right\} - \mu^2 = \sigma^2$ and $\mathrm{E}\left\{r_i r_j\right\} = \mathrm{E}\left\{r_i\right\}\mathrm{E}\left\{r_j\right\} = \mu^2$ for i.i.d. random variables. As it can be seen, $\mathrm{E}\left\{\hat{\sigma}_k^2\right\} \neq \sigma^2$. This means that the estimator according to Equation (3.6) is biased. However, since $\lim_{k\to\infty}\mathrm{E}\left\{\hat{\sigma}_k^2\right\} = \sigma^2$, this estimator is called asymptotically unbiased. An unbiased estimator for the variance is

$$\hat{\sigma}_k^2 = \frac{1}{k-1}\sum_{i=1}^{k}(r_i - \hat{\mu}_k)^2.$$

3.1.2   Cramér–Rao Lower Bound–CRLB

Since the observations r are noisy, it is interesting to evaluate the accuracy of an estimate $\hat{\alpha}(r)$. The variance $\mathrm{E}\left\{|\alpha - \hat{\alpha}(r)|^2\right\}$ of an estimate is a usual figure to quantify the accuracy of an estimator. In Example 3.1.1 we have calculated the variance of a mean value estimator (see Equation 3.4). The lower this variance, the higher is the accuracy of the estimator. This is also shown in Example 3.1.1, where we have lower bounded the probability of an estimate remaining within an interval of $\pm\epsilon$ around the true parameter value in Equation (3.5).

3.1.2.1   The Cauchy–Schwarz Inequality

It is straightforward to raise the question about the performance of an optimum estimator, that is, a lower bound on the variance of an estimate $\hat{\alpha}(r)$. Such a lower bound is provided by the Cramér–Rao inequality. The Cramér–Rao lower bound (CRLB) on the variance of an estimate is based on the Cauchy–Schwarz inequality.

Theorem 3.1.2 (Cauchy–Schwarz Inequality) *Let x and y be elements of an inner product space $\mathcal{V}$ over the field $\mathbb{F}$, where $\langle x, y \rangle : \mathcal{V} \times \mathcal{V} \mapsto \mathbb{F}$ denotes the inner product. $\mathbb{F}$ is either the field of real numbers $\mathbb{R}$ or complex numbers $\mathbb{C}$. Then $\forall x, y \in \mathcal{V}$,*

$$|\langle x, y \rangle|^2 \leq \langle x, x \rangle \langle y, y \rangle,$$

where $0 \leq |\lambda|^2 = \lambda \lambda^* \in \mathbb{R}$ denotes the squared absolute value of $\lambda \in \mathbb{F}$.

Proof. Using positive definiteness $0 \leq \langle x, x \rangle$, linearity $\langle \lambda x, x \rangle = \lambda \langle x, x \rangle$, $\langle x + y, z \rangle = \langle x, z \rangle + \langle y, z \rangle$ and conjugate symmetry $\langle x, y \rangle = \langle y, x \rangle^*$ of the inner product, we have

$$0 \leq \langle x - \lambda y, x - \lambda y \rangle = \langle x, x \rangle - \lambda^* \langle x, y \rangle - \lambda \langle x, y \rangle^* + |\lambda|^2 \langle y, y \rangle.$$

Setting $\lambda = \frac{\langle x, y \rangle}{\langle y, y \rangle}$, we get

$$0 \leq \langle x, x \rangle - \frac{\langle x, y \rangle^*}{\langle y, y \rangle^*} \langle x, y \rangle - \frac{\langle x, y \rangle}{\langle y, y \rangle} \langle x, y \rangle^* + \frac{|\langle x, y \rangle|^2}{|\langle y, y \rangle|^2} \langle y, y \rangle = \langle x, x \rangle - \frac{|\langle x, y \rangle|^2}{\langle y, y \rangle}.$$

Reordering this equation yields

$$|\langle x, y \rangle|^2 \leq \langle x, x \rangle \langle y, y \rangle$$

and proves Theorem 3.1.2.

Example 3.1.3 (Covariance) *Let us consider two random variables X and Y and their joint PDF $p_{XY}(x, y)$. We define the covariance of X and Y as*

$$\text{COV}\{X, Y\} = \langle X - \overline{X}, Y - \overline{Y} \rangle = E_{XY}\left\{(X - \overline{X})(Y - \overline{Y})^*\right\}$$

$$= \int \int (x - \overline{X})(y - \overline{Y})^* \, p_{XY}(x, y) \, dx \, dy$$

with $\overline{X} = E_X\{X\} = \int x \, p_X(x) \, dx$. According to Theorem 3.1.2, the absolute square of the covariance

$$|\text{COV}\{X, Y\}|^2 = \left|\langle X - \overline{X}, Y - \overline{Y} \rangle\right|^2$$

$$\leq \langle X - \overline{X}, X - \overline{X} \rangle \langle Y - \overline{Y}, Y - \overline{Y} \rangle = \text{VAR}\{X\}\,\text{VAR}\{Y\} \qquad (3.7)$$

can be upper bounded by the variances $\text{VAR}\{X\} = E_X\left\{|(X - \overline{X})|^2\right\}$ and $\text{VAR}\{Y\} = E_Y\left\{|(Y - \overline{Y})|^2\right\}$ of the random variables X and Y. Since

$$\langle X - \overline{X}, Y - \overline{Y} \rangle = \langle X, Y \rangle - \overline{X}\,\overline{Y}^*,$$

Equation (3.7) simplifies to

$$|\langle X, Y \rangle|^2 = \left|E_{XY}\{XY^*\}\right|^2 \leq \text{VAR}\{X\}\,\text{VAR}\{Y\} \qquad (3.8)$$

if either $\overline{X} = 0$ or $\overline{Y} = 0$.

3.1.2.2 Derivation of the Cramér–Rao Lower Bound

Before we start to derive the CRLB we look at the expectation value

$$E_{r|\alpha} \left\{ \frac{d}{d\alpha} \ln p(r|\alpha) \right\} = \int \left(\frac{d}{d\alpha} \ln p(r|\alpha) \right) p(r|\alpha)\, dr,$$

where $p(r|\alpha)$ is the conditional probability density function of a vector of observations (measurements) given a real valued parameter α, which we are interested in estimating. With $\frac{d}{d\alpha} \ln p(r|\alpha) = \frac{\frac{d}{d\alpha} p(r|\alpha)}{p(r|\alpha)}$ and exchanging integration and derivation, we get

$$E_{r|\alpha} \left\{ \frac{d}{d\alpha} \ln p(r|\alpha) \right\} = \frac{d}{d\alpha} \int p(r|\alpha)\, dr = \frac{d}{d\alpha} 1 = 0. \tag{3.9}$$

We are interested in the variance VAR $\{\hat{a}(r)\}$ of an estimator $\hat{a}(r)$. Thus, we investigate the covariance COV $\left\{ \hat{a}(r), \frac{d}{d\alpha} \ln p(r|\alpha) \right\}$. According to Equation (3.9), the expectation value of the second argument is zero. Due to Equation (3.8), the covariance simplifies. Then, we can write with the Cauchy–Schwarz Inequality as

$$\left| \text{COV} \left\{ \frac{d}{d\alpha} \ln p(r|\alpha), \hat{a}(r) \right\} \right|^2 = \left| E_{r|\alpha} \left\{ \left(\frac{d}{d\alpha} \ln p(r|\alpha) \right) \hat{a}(r) \right\} \right|^2$$

$$\leq E_{r|\alpha} \left\{ \left| \frac{d}{d\alpha} \ln p(r|\alpha) \right|^2 \right\} \text{VAR} \{\hat{a}(r)\}, \tag{3.10}$$

where we have used VAR $\left\{ \frac{d}{d\alpha} \ln p(r|\alpha) \right\} = E_{r|\alpha} \left\{ \left| \frac{d}{d\alpha} \ln p(r|\alpha) \right|^2 \right\}$, which follows from Equation (3.9). We have a closer look on the left-hand side of inequality Equation (3.10):

$$E_{r|\alpha} \left\{ \left(\frac{d}{d\alpha} \ln p(r|\alpha) \right) \hat{a}(r) \right\} = \int \left(\frac{d}{d\alpha} \ln p(r|\alpha) \right) \hat{a}(r) p(r|\alpha)\, dr$$

$$= \int \frac{\frac{d}{d\alpha} p(r|\alpha)}{p(r|\alpha)} \hat{a}(r)\, p(r|\alpha)\, dr = \frac{d}{d\alpha} \int \hat{a}(r)\, p(r|\alpha)\, dr = \frac{d}{d\alpha} E_{r|\alpha} \{\hat{a}(r)\}$$

Reordering Equation (3.10) yields the following theorem

Theorem 3.1.4 (Cramér–Rao Lower Bound) *The variance of an estimation $\hat{a}(r)$ of a parameter α based on observations r with conditional probability density function $p(r|\alpha)$ is lower bounded by*

$$\text{VAR} \{\hat{a}(r)\} \geq \frac{\left| \frac{d}{d\alpha} E_{r|\alpha} \{\hat{a}(r)\} \right|^2}{E_{r|\alpha} \left\{ \left| \frac{d}{d\alpha} \ln p(r|\alpha) \right|^2 \right\}}. \tag{3.11}$$

If the estimator is unbiased, that is, $E_{r|\alpha} \{\hat{a}(r)\} = \alpha$, Equation (3.11) simplifies to

$$\text{VAR} \{\hat{a}(r)\} \geq \frac{1}{E_{r|\alpha} \left\{ \left| \frac{d}{d\alpha} \ln p(r|\alpha) \right|^2 \right\}}. \tag{3.12}$$

Example 3.1.5 (CRLB of Gaussian mean value estimation) *We recall Example 3.1.1 and have a look on the mean value estimation there. The multivariate Gaussian distribution consisting of k i.i.d. scalar Gaussian random variables is*

$$p\left(r|\mu, \sigma^2\right) = \left(2\pi\sigma^2\right)^{-\frac{k}{2}} e^{-\frac{1}{2\sigma^2}\sum_{i=1}^{k}(r_i-\mu)^2},$$

where μ is the mean value of the components. First we calculate

$$E_{r|\mu}\left\{\left(\frac{d}{d\mu}\ln p(r|\mu)\right)^2\right\} = E_{r|\mu}\left\{\left(\frac{1}{2\sigma^2}\frac{d}{d\mu}\sum_{i=1}^{k}(r_i-\mu)^2\right)^2\right\}$$

$$= E_{r|\mu}\left\{\left(\frac{1}{2\sigma^2}\sum_{i=1}^{k}2(r_i-\mu)\right)^2\right\} = \frac{1}{\sigma^4}E_{r|\mu}\left\{\left(\sum_{i=1}^{k}(r_i-\mu)\right)^2\right\}$$

$$= \frac{1}{\sigma^4}\sum_{i=1}^{k}E_{r|\mu}\left\{(r_i-\mu)^2\right\} = \frac{k}{\sigma^2}.$$

The estimator in Equation (3.3) of Example 3.1.1 is unbiased. Therefore, we can apply Equation (3.12) and get

$$\text{VAR}\left\{\hat{\mu}(r)\right\} \geq \frac{1}{E_{r|\mu}\left\{\left(\frac{d}{d\mu}\ln p(r|\mu)\right)^2\right\}} = \frac{\sigma^2}{k}.$$

This means that there is no unbiased estimator providing a mean value estimation variance better than $\frac{\sigma^2}{k}$ for the Gaussian distribution. In Example 3.1.1 we have explicitly calculated the variance of the mean value estimation. Comparing this result (see Equation (3.4)) and the CRLB we have calculated previously, we can observe that the estimator we used in Example 3.1.1 is the optimum.

3.2 Propagation Time

Propagation time based positioning methods like TOA and TDOA require accurate estimation of the time of reception of signals coming from different transmitter sites. Similar to frame synchronization in wireless communications, this is a time estimation problem. The received signals are disturbed. A main source of such distortions is thermal noise, which is usually modeled as AWGN. Since multiple signals from different transmitters are received, interference may occur. The interference power itself depends on the cross-correlation properties of the used signals. In this section, we consider interference as a further additional Gaussian noise term. Because of noise, the estimation of the time of reception is erroneous as well. For repeated sinal transmission and time of reception estimation, this results in a jitter of the estimates around the true value. In this section we focus on timing estimation in the presence of AWGN.

3.2.1 Cramér–Rao Lower Bound for Time Estimation

Let's assume that we have transmitted a band limited continuous complex valued baseband signal $s(t)$. At the receiver, we obtain after sampling the discrete signal

$$r_\ell = s(\ell T_S - \tau) + n_\ell, \quad \ell = -L, \ldots, +L,$$

where n_ℓ denotes complex valued AWGN noise with zero mean and variance $E\left\{|n_\ell|^2\right\} = \sigma^2$. The received samples can be described by the multivariate Gaussian probability density function

$$p(r|\tau) = \left(\pi\sigma^2\right)^{-(2L+1)} e^{-\frac{1}{\sigma^2}\sum_{\ell=-L}^{L}|r_\ell - s(\ell T_S - \tau)|^2}.$$

According to Equation (3.12) and defining $s_\ell(\tau) := s(\ell T_S - \tau)$, we first calculate

$$\frac{d}{d\tau}\ln p(r|\tau) = -\frac{d}{d\tau}\frac{1}{\sigma^2}\sum_{\ell=-L}^{L}\left(r_\ell^R - s_\ell^R(\tau)\right)^2 + \left(r_\ell^I - s_\ell^I(\tau)\right)^2$$

$$= \frac{2}{\sigma^2}\sum_{\ell=-L}^{L}\left(r_\ell^R - s_\ell^R(\tau)\right)\frac{d}{d\tau}s_\ell^R(\tau) + \left(r_\ell^I - s_\ell^I(\tau)\right)\frac{d}{d\tau}s_\ell^I(\tau)$$

$$= \frac{2}{\sigma^2}\sum_{\ell=-L}^{L}n_\ell^R\frac{d}{d\tau}s_\ell^R(\tau) + n_\ell^I\frac{d}{d\tau}s_\ell^I(\tau),$$

where $(\cdot)^R$ and $(\cdot)^I$ denote real and imaginary part of a complex number. Taking the expectation over the absolute square results in

$$E\left\{\left|\frac{d}{d\tau}\ln p(r|\tau)\right|^2\right\} = \frac{4}{\sigma^4}\sum_{\ell=-L}^{L}\sum_{\ell'=-L}^{L}$$

$$E\left\{\left(n_\ell^R\frac{d}{d\tau}s_\ell^R(\tau) + n_\ell^I\frac{d}{d\tau}s_\ell^I(\tau)\right)\left(n_{\ell'}^R\frac{d}{d\tau}s_{\ell'}^R(\tau) + n_{\ell'}^I\frac{d}{d\tau}s_{\ell'}^I(\tau)\right)\right\}$$

$$= \frac{4}{\sigma^4}\sum_{\ell=-L}^{L}\underbrace{E\left\{(n_\ell^R)^2\right\}}_{=\frac{\sigma^2}{2}}\left(\frac{d}{d\tau}s_\ell^R(\tau)\right)^2 + \underbrace{E\left\{(n_\ell^I)^2\right\}}_{=\frac{\sigma^2}{2}}\left(\frac{d}{d\tau}s_\ell^I(\tau)\right)^2$$

$$= \frac{2}{\sigma^2}\sum_{\ell=-L}^{L}\left(\frac{d}{d\tau}s_\ell^R(\tau)\right)^2 + \left(\frac{d}{d\tau}s_\ell^I(\tau)\right)^2 = \frac{2}{\sigma^2}\sum_{\ell=-L}^{L}\left|\frac{d}{d\tau}s_\ell(\tau)\right|^2, \quad (3.13)$$

where we assume uncorrelated Gaussian noise, that is, $E\left\{n_\ell^R n_{\ell'}^I\right\} = 0$ and $E\left\{n_\ell^R n_{\ell'}^R\right\} = E\left\{n_\ell^I n_{\ell'}^I\right\} = \frac{\sigma^2}{2}\delta_{\ell\ell'}$. We insert the result form Equation (3.13) into Equation (3.12) and get for $L \to \infty$

$$\text{VAR}\{\hat{\tau}(r)\} \geq \frac{\sigma^2}{2\sum_{\ell=-\infty}^{\infty}\left|\frac{d}{d\tau}s(\ell T_S - \tau)\right|^2}. \quad (3.14)$$

It is obvious that the CRLB depends on the noise power. We observe from Equation (3.14), that this relationship is linear. If we increase the SNR by a factor of 10 (10 dB), the CRLB decreases by the same factor. Furthermore, signal properties influence the CRLB as well. We observe the sum over the absolute squares of the sampled derivative of $s(t)$. Thus, the more rapidly a signal changes, the lower the CRLB will be. Subsequently, we will clarify this relationship in more detail by describing the CRLB in terms of the spectrum of $s(t)$.

3.2.1.1 Frequency Domain Description

Previously, we have seen that rapid signal variation decreases the CRLB for timing estimation. Signal variation properties are easily visible by having a look at the spectrum of a signal. Using the Fourier transform, we describe the transmit signal

$$s(t) = \int_{-\infty}^{+\infty} S(f) e^{j2\pi ft} \, df$$

using its spectrum $S(f)$. We assume a bandlimited signal, that is, $S(f) = 0$ for $|f| > \frac{B}{2}$. Therefore, we can sample the continuous time signal using a sampling time of $T_S = \frac{1}{B}$ and get

$$s(\ell T_S) = \int_{-\frac{1}{2T_S}}^{+\frac{1}{2T_S}} S(f) e^{j2\pi f \ell T_S} \, df.$$

The inverse transformation is the time discrete Fourier series

$$\tilde{S}(f) = T_S \sum_{\ell=-\infty}^{+\infty} s(\ell T_S) e^{-j2\pi f \ell T_S}.$$

Note, $\tilde{S}(f) = \sum_{m=-\infty}^{+\infty} S(f - m B)$ is the periodic extension of $S(f)$. Due to band limitation, $S(f) = \tilde{S}(f)$ for $|f| \leq \frac{B}{2}$. According to Equation (3.14), we consider a time shift τ and get as the derivative

$$\frac{d}{d\tau} s(\ell T_S - \tau) = \frac{d}{d\tau} \int_{-\frac{1}{2T_S}}^{+\frac{1}{2T_S}} S(f) e^{j2\pi f(\ell T_S - \tau)} \, df$$

$$= \int_{-\frac{1}{2T_S}}^{+\frac{1}{2T_S}} \underbrace{-j2\pi f \, S(f) e^{-j2\pi f \tau}}_{\text{Fourier transform of } \frac{d}{d\tau} s(\ell T_S - \tau)} e^{j2\pi f \ell T_S} \, df. \tag{3.15}$$

Using Parseval's Theorem

$$\int_{-\frac{1}{2T_S}}^{+\frac{1}{2T_S}} |S(f)|^2 \, df = T_S \sum_{\ell=-\infty}^{+\infty} |s(\ell T_S)|^2$$

together with Equations (3.14) and (3.15) results in

$$\text{VAR}\left\{\hat{\tau}(r)\right\} \geq \frac{\sigma^2\, T_{\text{S}}}{8\pi^2 \int_{-\frac{1}{2T_{\text{S}}}}^{+\frac{1}{2T_{\text{S}}}} f^2\, |S(f)|^2\, \mathrm{d}f}. \tag{3.16}$$

Again, the dependency of VAR $\left\{\hat{\tau}(r)\right\}$ on the noise variance σ^2 is straightforward. A closer look at the denominator of Equation (3.16) shows that spending signal energy at higher frequencies enlarges the integral and yields better timing estimation performance. In the extreme case the entire signal energy is put into the maximum frequencies $f = \pm\frac{B}{2}$. This confirms the result of the previous section, where we have seen that rapid variations in the transmit signal $s(t)$ are beneficial for timing estimation.

3.2.1.2 Orthogonal Frequency Division Multiplexing–OFDM

In this section we will consider timing estimation for *orthogonal frequency division multiplexing (OFDM)*, which leads to another form of the CRLB in the frequency domain. OFDM is a multicarrier modulation scheme and deployed in many wireless broadband communications standards like WLAN, *Worldwide Interoperability for Microwave Access (WiMAX)* or 3GPP-LTE (see Chapter 8). In OFDM systems, the complex baseband transmit signal is usually designed in the frequency domain by complex valued symbols S_n, modulated on N subcarriers. The subcarrier spacing, that is, the difference in frequency of adjacent subcarriers, is chosen such that the subcarriers are orthogonal. Orthogonality holds for a subcarrier spacing of $f_{\text{SC}} = \frac{1}{T_{\text{OFDM}}}$, where T_{OFDM} is the duration of an OFDM symbol.

$$s(t) = \frac{1}{\sqrt{N}} \sum_{n=\lfloor -\frac{N-1}{2} \rfloor}^{\lfloor \frac{N-1}{2} \rfloor} S_n\, e^{j 2\pi\, n f_{\text{SC}} t}.$$

The delayed and sampled version of this signal can be obtained by an inverse discrete Fourier transform. After taking the derivative, we get the N time domain samples

$$\frac{\mathrm{d}}{\mathrm{d}\tau} s(\ell\, T_{\text{S}} - \tau) = \frac{1}{\sqrt{N}} \sum_{n=\lfloor -\frac{N-1}{2} \rfloor}^{\lfloor \frac{N-1}{2} \rfloor} -j\, 2\pi\, n f_{\text{SC}}\, S_n\, e^{j 2\pi\, n f_{\text{SC}}(\ell\, T_{\text{S}} - \tau)}$$

for $\ell = 0, \ldots, N-1$. Note that we have a time limited signal now. Starting from the denominator of Equation (3.14), we calculate

$$\sum_{\ell=-\infty}^{\infty} \left| \frac{\mathrm{d}}{\mathrm{d}\tau} s(\ell\, T_{\text{S}} - \tau) \right|^2 = \sum_{\ell=0}^{N-1} \frac{\mathrm{d}}{\mathrm{d}\tau} s(\ell\, T_{\text{S}} - \tau) \frac{\mathrm{d}}{\mathrm{d}\tau} s(\ell\, T_{\text{S}} - \tau)^*$$

$$= \frac{1}{N} \sum_{\ell=0}^{N-1} \sum_{m,n=\lfloor -\frac{N-1}{2} \rfloor}^{\lfloor \frac{N-1}{2} \rfloor} 4\pi^2\, mn f_{\text{SC}}^2\, S_m^* S_n\, e^{j 2\pi f_{\text{SC}}(\ell\, T_{\text{S}} - \tau)(n-m)}$$

$$= \frac{1}{N} \sum_{m,n=\lfloor -\frac{N-1}{2} \rfloor}^{\lfloor \frac{N-1}{2} \rfloor} 4\pi^2 \, mn f_{\text{SC}}^2 \, S_m^* S_n \underbrace{\sum_{\ell=0}^{N-1} e^{j 2\pi f_{\text{SC}}(\ell \, T_{\text{S}} - \tau)(n-m)}}_{=N \, \delta_{mn}}$$

$$= 4\pi^2 f_{\text{SC}}^2 \sum_{n=\lfloor -\frac{N-1}{2} \rfloor}^{\lfloor \frac{N-1}{2} \rfloor} n^2 \, |S_n|^2$$

Inserting this result into Equation (3.14) leads to

$$\text{VAR} \{\hat{\tau}(r)\} \geq \frac{\sigma^2}{8\pi^2 f_{\text{SC}}^2 \sum_{n=\lfloor -\frac{N-1}{2} \rfloor}^{\lfloor \frac{N-1}{2} \rfloor} n^2 \, |S_n|^2}. \tag{3.17}$$

For the case of OFDM, this result shows again the dependency of the CRLB on the spectral properties of the transmitted signal. Spending the available signal energy for the subcarriers at the spectrum edge maximizes the sum in the denominator or equivalently, provides the best timing estimation performance. We observe also that spending energy for subcarrier zero does not contribute improving timing estimation.

Usually, OFDM modulation is realized using an *inverse Fast Fourier transform (IFFT)*, which requires N to be a power of two. In order to keep the number of used subcarriers N_{u} flexible, zero padding is applied. This means that we do not use subcarriers in the range of $\frac{N_{\text{u}}}{2} \leq |n| < \frac{N}{2}$, that is, set $S_n = 0$ for these subcarriers. The application of Equation (3.17) is straightforward and will be shown in the next example.

Example 3.2.1 (CRLB for OFDM with uniformly distributed energy) *We assume an equal distribution of the signal energy*

$$|S_n|^2 = \begin{cases} |S|^2, & n \leq \frac{N}{2}, \\ 0, & \text{else} \end{cases}$$

among the used subcarriers $-M, \ldots, +M$, which are symmetrically grouped around subcarrier zero. The number of used subcarriers in this case is $N_{\text{u}} = 2M + 1$. Using Equation (3.17) leads to

$$\text{VAR} \{\hat{\tau}(r)\} \geq \frac{\sigma^2}{8\pi^2 f_{\text{SC}}^2 \, |S|^2 \sum_{n=-M}^{M} n^2} = \frac{1}{8\pi^2 f_{\text{SC}}^2 \frac{|S|^2}{\sigma^2} \frac{M(M+1)(2M+1)}{3}} \tag{3.18}$$

with an occupied bandwidth of

$$B = N_{\text{u}} f_{\text{SC}} = N_{\text{u}} \times 15 \, \text{kHz},$$

where $f_{\text{SC}} = 15 \, \text{kHz}$ is the subcarrier spacing of 3GPP-LTE (see Section 8.3). Figure 3.1 shows the CRLB for distance estimation in form of the standard deviation stddev $= c \sqrt{\text{VAR} \{\hat{\tau}(r)\}} = 3 \times 10^8 \, \text{m/s} \times \sqrt{\text{VAR} \{\hat{\tau}(r)\}}$ versus the signal bandwidth B for different receiver subcarrier signal-to-noise ratios (SNRs)

$$SNR = \frac{|S|^2}{\sigma^2}.$$

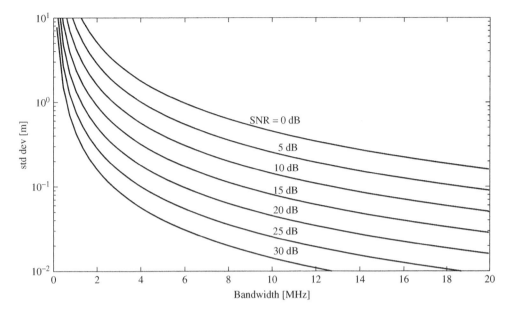

Figure 3.1   CRLB for TOA distance estimation

The strong dependency on the occupied bandwidth is clearly visible. Again, this indicates the value of high bandwidth signals for timing respectively ranging estimation.

3.2.2   Timing Estimation in White Gaussian Noise

We assume that L samples of a received signal in the complex baseband are available. The sample values

$$r_\ell = s(\ell\, T_\mathrm{S} - \tau) + n(\ell), \quad \ell = 0, \ldots, L-1,$$

consist of the transmitted signal $s(t)$, which undergoes a time delay τ and AWGN with zero mean and variance $\mathrm{E}\left\{ |n_\ell|^2 \right\} = \sigma^2$. It is straightforward to ask the question as to how we can extract an estimate of the signal delay τ from the measured signal r_ℓ. Thus, we have to find a metric, depending on τ, which we can use for optimization. An appropriate figure of merit is the *a posteriori* probability

$$\mathrm{p}(\tau|r) = \frac{\mathrm{p}(r|\tau)\,\mathrm{p}(\tau)}{\mathrm{p}(r)}.$$

Maximizing this probability leads to maximum a posteriori (MAP) estimation

$$\hat{\tau}(r) = \arg\max_\tau \mathrm{p}(\tau|r) = \arg\max_\tau \mathrm{p}(r|\tau)\,\mathrm{p}(\tau).$$

Instead of using probabilities we can also use logarithmic probabilities for maximization, that is,

$$\hat{\tau}(r) = \arg\max_\tau \ln \mathrm{p}(r|\tau) + \ln \mathrm{p}(\tau).$$

For applying the MAP rule we have to know the *a priori* probability $p(\tau)$. If this probability function is not known, a uniform distribution is typically assumed. With that assumption we arrive at the (logarithmic) maximum likelihood (ML) estimation rule

$$\hat{\tau}(r) = \arg \max_{\tau} \ln p(r|\tau), \tag{3.19}$$

which maximizes the likelihood function $p(r|\tau)$. For AWGN, the conditional PDF $p(r|\tau)$ is a multivariate Gaussian distribution

$$p(r|\tau) = \left(\pi\sigma^2\right)^{-L} e^{-\frac{1}{\sigma^2} \sum_{\ell=0}^{L-1} |r_\ell - s(\ell T_S - \tau)|^2}.$$

In this case, ML estimation is equivalent to minimizing the squared Euclidean distance between r and $s(\tau) = (s_0(\tau), \ldots, s_{L-1}(\tau))^T$ with $s_\ell(\tau) = s(\ell T_S - \tau)$, $\ell = 0, \ldots, L-1$.

$$\hat{\tau}(r) = \arg \min_{\tau} ||r - s(\tau)||_2^2 = \arg \min_{\tau} \sum_{\ell=0}^{L-1} |r_\ell - s(\ell T_S - \tau)|^2, \tag{3.20}$$

which is also called the mean squared error (MSE) estimation. Expanding the summand in Equation (3.20) yields

$$\hat{\tau}(r) = \arg \min_{\tau} \sum_{\ell=0}^{L-1} |r_\ell|^2 + |s(\ell T_S - \tau)|^2 - 2\operatorname{Re}\left\{r_\ell \, s(\ell T_S - \tau)^*\right\},$$

where $\operatorname{Re}\{\}$ denotes the real part of a complex number. Since $\sum_{\ell=0}^{L-1} |r_\ell|^2$ does not depend on τ, we can skip this term for minimization. If additionally $\sum_{\ell=0}^{L-1} |s(\ell T_S - \tau)|^2$ is constant with respect to τ, we can find the minimum in Equation (3.20) by maximizing the real part of the cross-correlation

$$\hat{\tau}(r) = \arg \max_{\tau} \operatorname{Re}\left\{\sum_{\ell=0}^{L-1} r_\ell \, s(\ell T_S - \tau)^*\right\}. \tag{3.21}$$

These signals are assumed to be complex valued baseband signals. At the transmitter, the signal $s(t)$ has to be upconverted into the RF band, where it is actually emitted. At the receiver, the RF signal must be downconverted into the baseband. This is typically done by using mixers and oscillators. The carrier phase and frequency of these oscillators are usually not known. In coherent receivers, these parameters are estimated and compensated such that the complex valued baseband signal at the receiver appears with correct phasing. Assuming an arbitrary but constant phase offset $\Delta\phi$ and an oscillator frequency offset Δf_O, the correlation sum in Equation (3.21) modifies to $e^{j\Delta\phi} \sum_{\ell=0}^{L-1} e^{j2\pi\Delta f_O \ell T_S} r_\ell \, s(\ell T_S - \tau)^*$. On the one hand, we observe that the phase offset shifts the phase of the correlation result. Using the real part for maximization as in Equation (3.21) is not beneficial if we do not know the phase offset. On the other hand the frequency offset term remains in the summand and degrades the correlation result. We can get rid of the constant phase term by using the absolute value of the correlation sum and obtain for non-coherent timing estimation

$$\hat{\tau}(r) = \arg \max_{\tau} \left|\sum_{\ell=0}^{L-1} r_\ell \, s(\ell T_S - \tau)^*\right|, \tag{3.22}$$

where we have neglected the carrier frequency offset. Because of building the absolute value, some noise terms, which are contained in r_ℓ appear quadratic. If the noise power is sufficiently low these terms can be neglected compared to the linear ones. However, at low SNRs these terms lead to a degradation, which is also called *squaring loss*.

3.2.2.1   Implementation: Delay Locked Loop–DLL

Subsequently we will introduce an implementation for timing estimation. We start from the ML estimation criterion in Equation (3.19) and find the minimum Euclidean distance between r_ℓ and $s(\ell\, T_S - \tau)$ with respect to τ by setting the first derivative of the log likelihood function to zero

$$\frac{d}{d\tau} \ln p(r|\tau) = -\frac{1}{\sigma^2}\frac{d}{d\tau}\sum_{\ell=0}^{L-1}|r_\ell - s(\ell\, T_S - \tau)|^2$$

$$= -\frac{1}{\sigma^2}\left(\underbrace{\frac{d}{d\tau}\sum_{\ell=0}^{L-1}|r_\ell|^2}_{=\frac{d}{d\tau}\,\text{const.}=0} + \frac{d}{d\tau}\sum_{\ell=0}^{L-1}|s(\ell\, T_S - \tau)|^2 - \frac{d}{d\tau}\sum_{\ell=0}^{L-1}2\,\mathrm{Re}\left\{r_\ell\, s(\ell\, T_S - \tau)^*\right\} \right)$$

$$= -\frac{1}{\sigma^2}\frac{d}{d\tau}\sum_{\ell=0}^{L-1}|s(\ell\, T_S - \tau)|^2 + \frac{2}{\sigma^2}\frac{d}{d\tau}\sum_{\ell=0}^{L-1}\mathrm{Re}\left\{r_\ell\, s(\ell\, T_S - \tau)^*\right\} \stackrel{!}{=} 0 \quad (3.23)$$

and then, solving for τ. For the derivation of Equation (3.21), we have assumed constant signal energy with respect to τ. In this case

$$\frac{d}{d\tau}\sum_{\ell=0}^{L-1}|s(\ell\, T_S - \tau)|^2 = \frac{d}{d\tau}\,\text{const.} = 0$$

and Equation (3.23) simplifies to

$$\mathrm{Re}\left\{\sum_{\ell=0}^{L-1} r_\ell \frac{d}{d\tau} s(\ell\, T_S - \tau)^*\right\} \stackrel{!}{=} 0, \quad (3.24)$$

which is the necessary condition for the maximum search in Equation (3.21). We can obtain the non-coherent version if we replace Re { } by | |. Equation (3.24) can be solved for τ iteratively using a closed control loop as shown in Figure 3.2. For implementation, the differential quotient in Equation (3.24) is approximated by a difference quotient with spacing Δ, that is

$$\frac{d}{d\tau}s(\ell\, T_S - \tau) \approx \frac{s\left(\ell\, T_S - \tau + \frac{\Delta}{2}\right) - s\left(\ell\, T_S - \tau - \frac{\Delta}{2}\right)}{\Delta}.$$

The loop filter usually contains an integral component. With that, it is guaranteed that the estimation $\hat{\tau}$ converges to the ML solution. This means that in the steady state the input signal to the loop filter is zero, which is the ML criterion according to Equation (3.24).

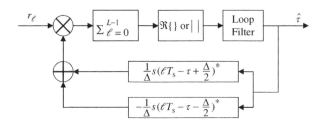

Figure 3.2   Delay locked loop (DLL)

3.2.2.2   Differential Timing Estimation

Previously, we have seen that timing estimation can be done by finding the maximum of a cross-correlation between the received signal and a replica of the transmitted signal, which of course must be known at the receiver. In contrast, differential correlation methods exploit knowledge about an appropriate signal structure instead of its full content information. Such a structure is signal symmetry-like repetitions, for instance. Let us consider a periodic structure of a signal of length L, beginning at time instance $\ell = 0$

$$s(\ell\, T_S) = s\left((\ell + L/2)\, T_S\right), \qquad \ell = 0, \dots \frac{L}{2} - 1,$$

which consists of two identical halves. Due to the properties of the discrete Fourier transform, such a signal can be constructed by setting every other subcarrier to zero, as is shown in Figure 3.3. Correlation of the received signal $r_\ell = s(\ell\, T_S - \tau) + n_\ell$, which is assumed to be corrupted by AWGN, with a replica of that signal, delayed by $L/2$ samples, provides an estimate of the signal delay τ (Schmidl and Cox 1997). The differential correlator output

$$D_k = \sum_{\ell=0}^{\frac{L}{2}-1} r_{\ell+k}\, r^*_{\ell+k+L/2} \tag{3.25}$$

is searched for the correlation peak

$$\hat{k}_0 = \arg\max_k \operatorname{Re}\left\{D_k\right\}.$$

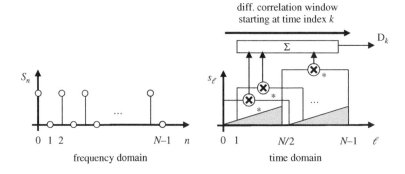

Figure 3.3   Differential correlation

This provides an estimation $\hat{\tau}(r) = k_0 T_S$ for the time delay τ. Figure 3.3 shows the principle of differential correlation. A correlation window is shifted over the received signal. If this window covers the two identical halves of the received signal, the correlation peak can be observed. Usually, in OFDM systems there is a guard interval in form of a cyclic prefix. In this case, the correlation function D_k shows a plateau-like peak (Schmidl and Cox 1997), which degrades the timing estimation performance. Increasing the number of signal periods as proposed for example, in (Minn et al. 2003) reduces this effect.

If carrier phase and frequency offsets are present in the system we observe the signal $\tilde{r}_\ell = e^{j\Delta\phi + j2\pi\Delta f_0 \ell T_S} r_\ell$. Equation (3.25) modifies to

$$\tilde{D}_k = \sum_{\ell=0}^{\frac{L}{2}-1} \tilde{r}_{\ell+k} \tilde{r}^*_{\ell+k+L/2}$$

$$= \sum_{\ell=0}^{\frac{L}{2}-1} r_{\ell+k} e^{j\Delta\phi} e^{j2\pi\Delta f_0(\ell+k)T_S} r^*_{\ell+k+L/2} e^{-j\Delta\phi} e^{-j2\pi\Delta f_0(\ell+k+L/2)T_S}$$

$$= e^{-j2\pi\Delta f_0 L/2 T_S} \sum_{\ell=0}^{\frac{L}{2}-1} r_{\ell+k} r^*_{\ell+k+L/2}$$

$$= e^{-j2\pi\Delta f_0 L/2 T_S} D_k.$$

We observe that a carrier phase offset cancels out. A carrier frequency offset, however, remains as a constant phase rotation of the correlation result. We can get rid of this phase factor by using the absolute value of the correlation sum for timing estimation

$$\hat{k}_0 = \arg \max_k |\tilde{D}_k| = \arg \max_k |D_k|.$$

instead of its real part.

3.2.2.3 Reverse Differential Timing Estimation

Using OFDM, there is another approach for generation of signal symmetry in the time domain. These kind of signals are generated from real valued symbols in the frequency domain. The time domain signal s_k, $k = 0, \ldots, N-1$ is obtained by applying an inverse discrete Fourier transform (DFT) to the respective real valued frequency domain signal S_n, $n = 0, \ldots, N-1$. Using real valued frequency domain symbols results in $\text{Im}\{S_n\} = 0$ or consequently

$$S_n = S_n^*.$$

To show the resulting symmetry, we start from the definition of the inverse DFT at time instance $N - \ell$ and get

$$s_{N-\ell} = \frac{1}{\sqrt{N}} \sum_{n=0}^{N-1} S_n e^{j\frac{2\pi}{N}n(N-\ell)} = \frac{1}{\sqrt{N}} \sum_{n=0}^{N-1} S_n \underbrace{e^{j2\pi n}}_{=1} e^{-j\frac{2\pi}{N}n\ell}$$

$$= \frac{1}{\sqrt{N}} \left(\sum_{n=0}^{N-1} \underbrace{S_n^*}_{=S_n} e^{j\frac{2\pi}{N}n\ell} \right)^* = s_\ell^*.$$

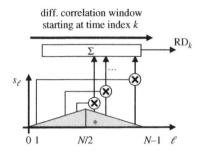

Figure 3.4   Reverse differential correlation

This means the second half is the conjugate complex mirror image of the first half of the signal. Figure 3.4 shows that symmetry property. For this kind of symmetry, we have to apply reverse differential correlation (Berggren and Popovic 2007), that is, we have to correlate the received signal with the reverse image of the succeeding interval as shown in Figure 3.4. Since the reverse image is already conjugate complex, conjugation of the received signal, as for differential correlation, is not required.

$$RD_k = \sum_{\ell=1}^{\frac{N}{2}-1} r_{k+\ell}\, r_{N-\ell+k}. \tag{3.26}$$

Again, an estimate for signal delay τ is obtained by searching for the correlation peak. For coherent timing estimation we get

$$\hat{k}_0 = \arg\max_k \mathrm{Re}\left\{RD_k\right\}.$$

Receiving a signal $\tilde{r}_\ell = e^{j\Delta\phi+j2\pi\Delta f_0\ell\,T_S}r_\ell$, which includes carrier phase and frequency offsets leads to

$$\tilde{RD}_k = \sum_{\ell=1}^{\frac{N}{2}-1} \tilde{r}_{k+\ell}\, \tilde{r}_{N-\ell+k}$$

$$= \sum_{\ell=1}^{\frac{N}{2}-1} e^{j\Delta\phi+j2\pi\Delta f_0 T_S\,(k+\ell)}\, r_{k+\ell}\, e^{j\Delta\phi+j2\pi\Delta f_0 T_S\,(N-\ell+k)}\, r_{N-\ell+k}$$

$$= e^{j\Delta\phi+j2\pi\Delta f_0 T_S\,(2k+N)}\, RD_k.$$

We observe a phase shift of the correlation result compared to coherent correlation in Equation (3.26). As for differential correlation, we may use the non-coherent estimation method

$$\hat{k}_0 = \arg\max_k \left|\tilde{RD}_k\right| = \arg\max_k \left|RD_k\right|.$$

For both differential correlation methods, the time delay estimate is quantized in intervals of the sampling time T_S. To overcome this restriction, we may apply interpolation of the received signal. This could be achieved by steering the phase of the signal sampling unit at the receiver front-end in a closed loop, including the correlation entity. Interpolation can be implemented in the frequency domain. Using the discrete Fourier transform an arbitrary signal time shift corresponds to a phase shift of the frequency domain symbols with linearly increasing phase. We will introduce this method subsequently.

3.2.2.4   Timing Estimation in the Frequency Domain

In OFDM, signal processing is typically applied in the frequency domain. At the transmitter signals are defined and multiplexed in the frequency domain before being transformed into the time domain using an inverse DFT. At the receiver, this is reverted. Time domain samples are transformed into the frequency domain where subsequent signal processing like channel estimation and data symbol detection is performed. Previously, we have seen that timing estimation in the time domain is based on sampled signals. Therefore, the estimation accuracy is quantized to the sampling time grid. In 3GPP-LTE, the subcarrier spacing is $f_{SC} = 15\,\text{kHz}$ (see Section 8.3 for more details). Applying an FFT-Length of $N = 2048$, this leads to a sampling time of $T_S = \frac{1}{N f_{SC}} = 32.55\,\text{ns}$. For communications using OFDM, an accuracy quantized to that sampling time is sufficient. Slight inaccuracies are covered by the guard interval and compensated by channel estimation and equalization. However, for localization purposes, an estimation error of one time sample means a ranging error of $c\,T_S = 32.55\,\text{ns} \cdot 3 \cdot 10^{-8}\,\text{m/s} = 9.77\,\text{m}$, which is not acceptable for many applications. Thus, we need non-discrete timing estimation methods, which means that we have to apply interpolation. This can be seen for example, from the correlation sum in Equations (3.21) or (3.22), where we assume delay parameter $\tau \in \mathbb{R}$, that is, this parameter is continuous rather than discrete. In OFDM, signal processing is mainly based on data samples in the frequency domain, which makes interpolation easily applicable. To show this, we start at the definition of the time discrete Fourier transform, where we can obtain a continuous time signal in the complex base band by

$$s(t) = \frac{1}{\sqrt{N}} \sum_{n=\lfloor -\frac{N-1}{2}\rfloor}^{\lfloor \frac{N-1}{2}\rfloor} S_n\, e^{j2\pi n f_{SC} t}.$$

Sampling at $t = \ell\, T_S = \ell\, \frac{1}{N f_{SC}}$ yields a definition of the inverse DFT. Setting $t = (\ell\, T_S - \tau)$ leads to

$$s(\ell\, T_S - \tau) = \frac{1}{\sqrt{N}} \sum_{n=\lfloor -\frac{N-1}{2}\rfloor}^{\lfloor \frac{N-1}{2}\rfloor} S_n\, e^{j2\pi n f_{SC}\, (\ell\, T_S - \tau)}$$

$$= \frac{1}{\sqrt{N}} \sum_{n=\lfloor -\frac{N-1}{2}\rfloor}^{\lfloor \frac{N-1}{2}\rfloor} \underbrace{e^{-j2\pi n f_{SC}\, \tau}\, S_n}_{\text{DFT of } s(\ell\, T_S - \tau)}\, e^{j\frac{2\pi}{N} n \ell}.$$

$$(3.27)$$

Thus, a continuous time shift corresponds to a linearly increasing phase shift of the signal symbols in the frequency domain. Using the DFT property in Equation (3.27) we can express the correlation sum in Equation (3.21) as

$$\sum_{\ell=0}^{N-1} r_\ell \, s(\ell \, T_S - \tau)^* = \sum_{\ell=0}^{N-1} \frac{1}{\sqrt{N}} \sum_{m=\lfloor -\frac{N-1}{2}\rfloor}^{\lfloor \frac{N-1}{2}\rfloor} R_m \, e^{j\frac{2\pi}{N}m\ell} \frac{1}{\sqrt{N}} \sum_{n=\lfloor -\frac{N-1}{2}\rfloor}^{\lfloor \frac{N-1}{2}\rfloor} e^{j2\pi nf_{SC}\,\tau} \, S_n^* \, e^{-j\frac{2\pi}{N}n\ell}$$

$$= \frac{1}{N} \sum_{m=\lfloor -\frac{N-1}{2}\rfloor}^{\lfloor \frac{N-1}{2}\rfloor} \sum_{n=\lfloor -\frac{N-1}{2}\rfloor}^{\lfloor \frac{N-1}{2}\rfloor} R_m \, e^{j2\pi nf_{SC}\,\tau} \, S_n^* \underbrace{\sum_{\ell=0}^{N-1} e^{j\frac{2\pi}{N}(m-n)\ell}}_{=N\,\delta_{mn}}$$

$$= \sum_{n=\lfloor -\frac{N-1}{2}\rfloor}^{\lfloor \frac{N-1}{2}\rfloor} e^{j2\pi nf_{SC}\,\tau} \, R_n \, S_n^*.$$

With this result we can rewrite coherent timing estimation in the frequency domain as

$$\hat{\tau}(r) = \arg \max_\tau \, \mathrm{Re}\left\{ \sum_{n=\lfloor -\frac{N-1}{2}\rfloor}^{\lfloor \frac{N-1}{2}\rfloor} e^{j2\pi nf_{SC}\,\tau} \, R_n \, S_n^* \right\}. \tag{3.28}$$

For the non-coherent case, we get

$$\hat{\tau}(r) = \arg \max_\tau \, \left| \sum_{n=\lfloor -\frac{N-1}{2}\rfloor}^{\lfloor \frac{N-1}{2}\rfloor} e^{j2\pi nf_{SC}\,\tau} \, R_n \, S_n^* \right|. \tag{3.29}$$

Example 3.2.2 (Synchronization in 3GPP-LTE) *The frame structure of 3GPP-LTE comprises sequences, which are used for timing, carrier phase, and frequency offset estimation or even Cell-ID detection (see Section 8.3). These signals are called* Primary Synchronization Sequences (PSSs) *and* Secondary Synchronization Sequences (SSSs). *They have a length of 62 symbols and occupy subcarriers* $n = -31, -30, \ldots, -1, +1, +2, \ldots, +31$. *In particular, the subcarrier at frequency zero is not used. We continue Example 3.2.1 and calculate the CRLB for ranging performance versus the subcarrier SNR*

$$\mathrm{SNR} = \frac{|S|^2}{\sigma^2}$$

using Equation (3.18), where $N_u = 2M + 1 = 2 \times 31 + 1 = 63$. *With a subcarrier spacing of* $f_{SC} = 15 \, \mathrm{kHz}$, *the 3GPP-LTE synchronization signals occupy a bandwidth of*

$$B = N_u f_{SC} = 63 \times 15 \, \mathrm{kHz} = 945 \, \mathrm{kHz}.$$

We compare this bound with simulation results according to coherent and non-coherent estimation according to Equations (3.28) and (3.29). Figure 3.5 shows this comparison. It can be seen that both coherent and non-coherent timing estimation reach the bound for a wide range of SNRs. At very low SNRs, the estimation performance diverges from the bound. Additionally, a performance difference between coherent and non-coherent estimation is observable.

3.3   Angle of Arrival–AOA

The AOA of an incident wavefront is another quantity that can be used to estimate the direction of the transmitter position. As already pointed out in Section 2.2.4 we have to assume LOS propagation for positioning. Propagation effects like diffraction, refraction, or reflection change the direction of a wavefront, leading to wrong position estimates. However, for this section we focus on AOA estimation. For that, we need signal measures that depend on the direction of the incident wavefront. Directional antennas provide such dependencies. Subsequently, we focus on antenna arrays, which allow beamforming by appropriate digital signal processing.

3.3.1   Uniform Linear Array Antenna

We assume a set of L identical antenna elements in a linear array as shown in Figure 3.6. The distance between adjacent elements is d. A planar wavefront, received from an angle α, shows an incremental signal traveling distance of

$$\Delta = d \, \sin(\alpha)$$

until it reaches the different antenna elements indexed by $\ell = 0, \dots, L-1$. Assuming a sinusoidal waveform, we observe a linearly increasing phase of

$$\phi_\ell = -\left(\ell - \frac{L-1}{2}\right) kd \, \sin(\alpha),$$

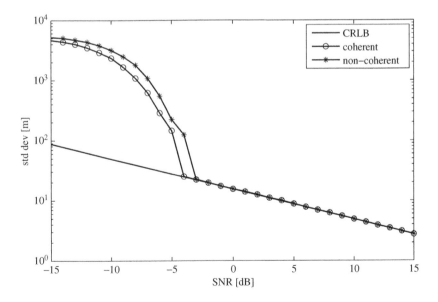

Figure 3.5   TOA distance estimation performance

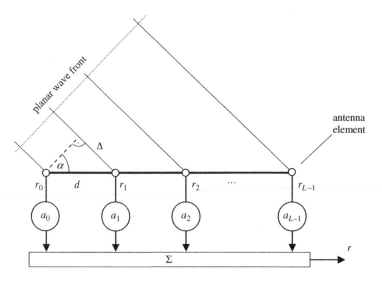

Figure 3.6 Uniform linear array antenna

where $k = \frac{2\pi}{\lambda}$ is the wavenumber and $\lambda = c/f_0$ is the wavelength that can be calculated from the speed of light c and the carrier frequency f_0. The phase is chosen such that it is symmetric with respect to the antenna array. Neglecting noise for the moment, we receive a signal

$$r_\ell = s\, e^{-j\left(\ell-\frac{L-1}{2}\right)kd\,\sin(\alpha)} \tag{3.30}$$

in complex baseband notation at antenna element ℓ. s denotes the amplitude of the received signal. Since the additional signal traveling distances $\left(\ell - \frac{L-1}{2}\right)\Delta$, $\ell = 0, \dots, L-1$ are very small compared to the distance between transmitter and receiver, it is reasonable to assume that s is constant for all signals r_ℓ received at antenna elements $\ell = 0, \dots, L-1$. The received signals r_n are weighted by amplitude factors a_ℓ and added. We may use these weighting factors for compensation of the phase differences ϕ_ℓ in order to achieve constructive addition of the received signals induced by the incident wavefront from a certain direction α_0. In other words, with

$$a_\ell = e^{j\theta_\ell} = e^{j\left(\ell-\frac{L-1}{2}\right)kd\,\sin(\alpha_0)}$$

we steer the main lobe of the antenna array beam into direction α_0. The vector $a = [a_0, a_1, \dots, a_{L-1}]$ is called the steering vector. With this choice of a_ℓ, the superimposed signal is

$$r = \sum_{\ell=0}^{L-1} a_\ell\, r_\ell = \sum_{\ell=0}^{L-1} s\, e^{-j(\phi_\ell - \theta_\ell)}$$

$$= s\, e^{j\frac{L-1}{2}kd\,(\sin(\alpha)-\sin(\alpha_0))} \sum_{\ell=0}^{L-1} e^{-j\ell kd\,(\sin(\alpha)-\sin(\alpha_0))}. \tag{3.31}$$

It is easy to show that we have constructive addition for $\alpha = \alpha_0$. Due to symmetry of the antenna array, constructive superposition also happens for an AOA of $\alpha = \pi - \alpha_0$. We focus on the AOA interval $\alpha = [-\pi/2, +\pi/2)$ and assume that wavefronts coming from directions outside of that interval, that is, from below the linear array in Figure 3.6, are shielded.

The sum term in Equation (3.31) is a finite geometric sum of the form $\sum_{\ell=0}^{L-1} q^\ell = \frac{1-q^L}{1-q}$ with $q = e^{-jkd\,(\sin(\alpha)-\sin(\alpha_0))}$. Thus, we can rewrite Equation (3.31) as

$$r = s\,e^{j\frac{L-1}{2}kd\,(\sin(\alpha)-\sin(\alpha_0))}\,\frac{1 - e^{-jLkd\,(\sin(\alpha)-\sin(\alpha_0))}}{1 - e^{-jkd\,(\sin(\alpha)-\sin(\alpha_0))}}$$

$$= s\,\frac{\sin\left[L\frac{\pi d}{\lambda}\,(\sin(\alpha) - \sin(\alpha_0))\right]}{\sin\left[\frac{\pi d}{\lambda}\,(\sin(\alpha) - \sin(\alpha_0))\right]}. \tag{3.32}$$

The amplitude of Equation (3.32) is of form $D_L(x) = \left|\frac{\sin(Lx)}{\sin(x)}\right|$. Maxima appear at at $x = n\pi$, $n \in \mathbb{Z}$. So, the amplitude has a periodicity of π. Later, we will use the amplitude part of Equation (3.32) for AOA estimation by searching for the maximum in the interval $\alpha_0 = [-\pi/2, \pi/2)$. In order to observe only one maximum, we have to restrict the argument of $D_L(x)$ to one period, that is, $|x| < \pi$. Since $|\sin(\alpha) - \sin(\alpha_0)| \leq 2$ we have $|x| = \left|\frac{\pi d}{\lambda}(\sin(\alpha) - \sin(\alpha_0))\right| \leq \frac{2\pi d}{\lambda} < \pi$ or equivalently

$$d < \frac{\lambda}{2}. \tag{3.33}$$

According to Equation (3.33), the antenna element spacing d has to be smaller than $\lambda/2$ in order to avoid multiple maxima. Choosing the spacing d significantly smaller than $\lambda/2$ results in a reduced aperture, which causes a reduced angular resolution.

Equation (3.32) is the array factor of an uniform linear antenna array. The entire antenna array characteristic

$$A(\alpha) = AF(\alpha) \times EF(\alpha)$$

can be obtained by multiplication of the array factor $AF(\alpha)$ and the element factor $EF(\alpha)$. By properly choosing the steering vector a, we can influence the array factor. However, the element factor is a property of the used antenna elements and cannot be influenced adaptively.

For comparison, we have normalized the group factor for a maximum value of 1, regardless of the number of antenna elements. Assuming a signal amplitude of $s = 1$, we obtain the amplitude factor from the amplitude of the normalized array factor

$$|AF(\alpha)| = \left|\frac{\sin\left[L\frac{\pi d}{\lambda}\,(\sin(\alpha) - \sin(\alpha_0))\right]}{L\sin\left[\frac{\pi d}{\lambda}\,(\sin(\alpha) - \sin(\alpha_0))\right]}\right|.$$

In Figure 3.7, we have plotted $|AF(\alpha)|$ for different antenna array parameters. Figures 3.7(a)–(c) show the array factor for different antenna beam directions. We observe that the width of the main lobe, that is, the angular region around the maximum of $|AF(\alpha)|$, increases with an increasing angle α_0. This is due to the fact that with an increasing angle α_0 the visible antenna aperture decreases. The visible aperture, that is the observable 'size' of the antenna from direction α_0, is proportional to $\cos(\alpha_0)$. A broader main lobe results in reduced AOA estimation performance. Thus, the variance of an AOA estimator is not constant over the angular

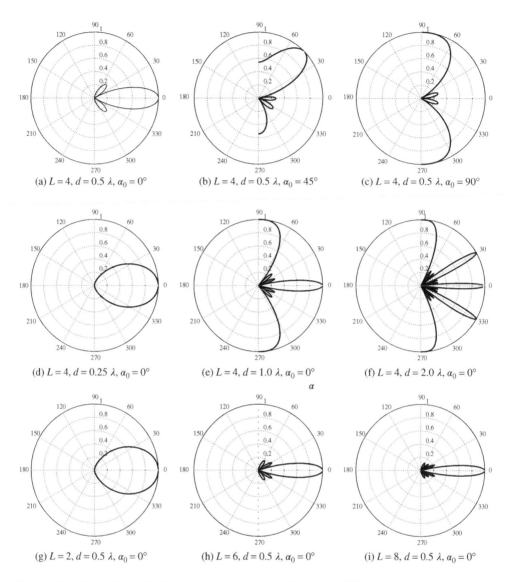

Figure 3.7 Comparison of uniform linear antenna array factors for different numbers of elements L, antenna element spacings d and AOAs α_0

range $\alpha = [-\pi/2, +\pi/2)$. Figures 3.7(d)–(f) show the dependency of the array factor on the antenna element spacing d. As we increase d beyond $\lambda/2$, we observe multiple main lobes. This results in AOA estimation ambiguity. On the other hand, with an increasing element spacing d, the antenna aperture increases as well. This results in a reduced main lobe width. Finally, Figures 3.7(g)–(i) show the array factor for a different number of antenna elements L. Since the antenna aperture is proportional to L, the main lobe width decreases with an increasing number of antenna elements. Between the periodic occurring main lobes of $|AF(\alpha)|$, there are

$L - 1$ roots. Between these roots, there are $L - 2$ side lobes, that is, local maxima of $|AF(\alpha)|$. Comparing Figures 3.7(d) and (g), we observe a similar main lobe width, which is a result of similar antenna array apertures $L \times d = 4 \times \lambda/4 = 2 \times \lambda/2 = \lambda$. The same holds for the arrays shown in Figures 3.7(e) and (i) with an aperture of $L \times d = 4\lambda$.

3.3.2   AOA Estimation in Additive White Gaussian Noise

In Equation (3.30), we have already introduced a description for the received signals at antenna element ℓ of an uniform linear array antenna, where we have assumed a planar incident wavefront. Besides the useful signal, each antenna element also receives electromagnetic radiation from the environment. We model this as AWGN and thus, obtain the received signal

$$r_\ell = s\,e^{j\left(\psi - \left(\ell - \frac{L-1}{2}\right)kd\sin(\alpha)\right)} + n_\ell, \tag{3.34}$$

where $s \geq 0 \in \mathbb{R}$ is the amplitude of the received signal induced by the incident wavefront. ψ is an arbitrary phase, which is constant for all antenna elements. n_ℓ denotes complex valued AWGN with zero mean and variance σ^2. Using this signal model, we can write the likelihood function as

$$p(r|\alpha) = \left(\pi\sigma^2\right)^{-L} \exp\left(-\frac{1}{\sigma^2}\sum_{\ell=0}^{L-1}\left|r_\ell - s\,e^{j\left(\psi - \left(\ell - \frac{L-1}{2}\right)kd\sin(\alpha)\right)}\right|^2\right). \tag{3.35}$$

For the ML estimation, we have to minimize the squared Euclidean distance between r and $s(\alpha)$ with $s_\ell(\alpha) = s\,e^{j\left(\psi - \left(\ell - \frac{L-1}{2}\right)kd\sin(\alpha)\right)}$, $\ell = 0, \ldots, L - 1$, that is,

$$\hat{\alpha}(r) = \arg\min_\alpha \sum_{\ell=0}^{L-1}\left|r_\ell - s\,e^{j\left(\psi - \left(\ell - \frac{L-1}{2}\right)kd\sin(\alpha)\right)}\right|^2, \tag{3.36}$$

Expanding the summand in Equation (3.36) yields

$$\hat{\alpha}(r) = \arg\min_\alpha \sum_{\ell=0}^{L-1}\left(\left|r_\ell\right|^2 + \left|s\,e^{j\left(\psi - \left(\ell - \frac{L-1}{2}\right)kd\sin(\alpha)\right)}\right|^2\right.$$
$$\left. -2\,\mathrm{Re}\left\{r_\ell\,s\,e^{-j\left(\psi - \left(\ell - \frac{L-1}{2}\right)kd\sin(\alpha)\right)}\right\}\right), \tag{3.37}$$

where $\mathrm{Re}\{\}$ denotes the real part of a complex number. Both $\sum_{\ell=0}^{L-1}\left|r_\ell\right|^2$ and $\sum_{\ell=0}^{L-1}\left|s\,e^{j\left(\psi - \left(\ell - \frac{L-1}{2}\right)kd\sin(\alpha)\right)}\right|^2$ are constant with respect to α. Thus, we can find the minimum in Equation (3.37) by maximizing the real part of the cross-correlation and get

$$\hat{\alpha}(r) = \arg\max_\alpha \mathrm{Re}\left\{e^{-j\psi}\sum_{\ell=0}^{L-1}r_\ell\,e^{j\left(\ell - \frac{L-1}{2}\right)kd\sin(\alpha)}\right\}. \tag{3.38}$$

for AOA estimation. Equation (3.38) does not depend on s since the signal amplitude is assumed to be a real valued positive number. To solve Equation (3.38), we have to know the initial phase ψ. In a coherent approach this phase may be estimated and compensated. We can avoid this by using the absolute value instead of building the real part in Equation (3.38). This leads to non-coherent AOA estimation

$$\hat{\alpha}(r) = \arg \max_{\alpha} \left| \sum_{\ell=0}^{L-1} r_\ell \, e^{j\left(\ell - \frac{L-1}{2}\right)kd\sin(\alpha)} \right|.$$

3.3.2.1 AOA Estimation using Linear Regression

AOA estimation, as introduced previously, requires the maximization of the likelihood function. We observe from the received signals at antenna element $\ell = 1, \dots, L-1$

$$r_\ell = s \, e^{j\left(\psi - \left(\ell - \frac{L-1}{2}\right)kd\sin(\alpha)\right)} + n_\ell$$

that the phase is linear in ℓ. We will use the argument, that is, the phase values, of the noisy samples r_ℓ for calculation of a regression line. Then, the steepness of this line is a minimum mean squared error estimation for $-kd\sin(\alpha)$.

The argument of the complex value r_ℓ

$$\rho_\ell = \arg(r_\ell) = \psi - \tilde{\ell} \, kd\sin(\alpha) + v_\ell$$

consists of the argument $\phi_\ell = \psi - \tilde{\ell} \, kd\sin(\alpha)$ of the wanted signal and a phase noise term v_ℓ. For notational convenience we use $\tilde{\ell} = \left(\ell - \frac{L-1}{2}\right)$. Figure 3.8 illustrates the transition of AWGN in the complex signal domain into phase noise. The complex valued AWGN n with zero mean and variance σ^2 is separated into a noise component $n_{\parallel}$, which is parallel to the wanted signal, and an orthogonal component $n_{\perp}$. Both $n_{\parallel}$ and $n_{\perp}$ have variance $\sigma^2/2$. A phase noise term can be approximated by

$$v_\ell = \arctan\left(\frac{n_{\perp}}{s + n_{\parallel}}\right) \approx \arctan\left(\frac{n_{\perp}}{s}\right) \approx \frac{n_{\perp}}{s}$$

if the SNR is sufficiently high, that is, $s \gg |n|$. The second approximation is due to $y = \tan(x) \approx x$ for $x \ll 1$. With these approximations, the phase noise term ϕ_n is real valued additive white Gaussian noise with zero mean and variance

$$E\left\{v_\ell^2\right\} \approx \frac{\sigma^2}{2s^2} = \frac{1}{2\,\mathrm{SNR}},$$

where $\mathrm{SNR} = \frac{s^2}{\sigma^2}$ is the SNR in the signal domain.

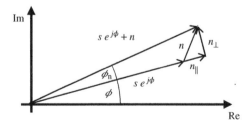

Figure 3.8   Phase noise approximation

The linear regression model is

$$\phi_\ell = \psi - \tilde{\ell}\, kd \sin(\alpha) = \psi + \beta\,\tilde{\ell}.$$

From the L signal samples r_ℓ, we calculate the phase samples $\rho_\ell = \arg(r_\ell)$. The linear regression model parameters are

$$-kd \sin(\hat{\alpha}) = \beta = \frac{\sum_{\ell=0}^{L-1}(\tilde{\ell} - \overline{\tilde{\ell}})(\rho_\ell - \overline{\rho})}{\sum_{\ell=0}^{L-1}(\tilde{\ell} - \overline{\tilde{\ell}})^2} = \frac{12\sum_{\ell=0}^{L-1}\tilde{\ell}\,(\rho_\ell - \overline{\rho})}{(L-1)L(L+1)} \tag{3.39}$$

$$\psi = \overline{\rho}_\ell - \beta\,\overline{\tilde{\ell}} = \overline{\rho},$$

where $\overline{\tilde{\ell}} = \frac{1}{L}\sum_{\ell=0}^{L-1}\tilde{\ell} = 0$ and $\overline{\rho} = \frac{1}{L}\sum_{\ell=0}^{L-1}\rho_\ell$ are the sample mean values. To obtain the last equality in Equation (3.39), we use

$$\sum_{\ell=0}^{L-1}\tilde{\ell}^2 = \sum_{\ell=0}^{L-1}\left(\ell - \frac{L-1}{2}\right)^2 = \frac{L(L^2-1)}{12}. \tag{3.40}$$

The estimate for the incident waveform direction is then given by

$$\hat{\alpha} = \arcsin\left(-\frac{\beta}{kd}\right)$$

in radians.

3.3.3   Cramér–Rao Lower Bound for AOA Estimation

To derive the CRLB for AOA estimation, we use the RX signal model and the likelihood function defined by Equations 3.34 and 3.35 in Section 3.3.2. For notational convenience, we introduce $\tilde{\ell} = \left(\ell - \frac{L-1}{2}\right)$. We recall Equation (3.12), where we first calculate

$$\frac{d}{d\alpha}\ln p(r|\alpha) = -\frac{d}{d\alpha}\frac{1}{\sigma^2}\sum_{\ell=0}^{L-1}\left|r_\ell - s\,e^{j(\psi-\tilde{\ell}kd\sin(\alpha))}\right|^2$$

$$= \frac{1}{\sigma^2}\sum_{\ell=0}^{L-1}2\left[r_\ell^R - s\cos\left(\psi - \tilde{\ell}kd\sin(\alpha)\right)\right]\frac{d}{d\alpha}s\cos\left(\psi - \tilde{\ell}kd\sin(\alpha)\right)$$

$$+ 2\left[r_\ell^I - s\sin\left(\psi - \tilde{\ell}kd\sin(\alpha)\right)\right]\frac{d}{d\alpha}s\sin\left(\psi - \tilde{\ell}kd\sin(\alpha)\right)$$

$$= \frac{2\,s\,kd\cos(\alpha)}{\sigma^2}\sum_{\ell=0}^{L-1}\operatorname{Re}\left\{n_\ell\right\}\tilde{\ell}\sin\left(\psi - \tilde{\ell}kd\sin(\alpha)\right) - \operatorname{Im}\left\{n_\ell\right\}\tilde{\ell}\cos\left(\psi - \tilde{\ell}kd\sin(\alpha)\right).$$

Taking the expectation over the absolute square results in

$$E\left\{\left|\frac{d}{d\alpha}\ln p(r|\alpha)\right|^2\right\} = \frac{2\,s^2\,k^2d^2\cos^2(\alpha)}{\sigma^2}\sum_{\ell=0}^{L-1}\tilde{\ell}^2 = \frac{L(L^2-1)\,s^2\,k^2d^2\cos^2(\alpha)}{6\sigma^2}$$

where we assume uncorrelated Gaussian noise, that is $E\{Re\{n_\ell\}\,Im\{n_\ell\}\} = 0$ and $E\{Re\{n_\ell\}\,Re\{n_\ell\}\} = E\{Im\{n_\ell\}\,Im\{n_\ell\}\} = \frac{\sigma^2}{2}\delta_{\ell\ell'}$. We insert this result into Equation (3.12) and get

$$VAR\{\hat{a}(r)\} \geq \frac{6}{L(L^2-1)\,SNR\,k^2d^2\cos^2(\alpha)}, \qquad (3.41)$$

where we have used Equation (3.40) and the $SNR = \frac{s^2}{\sigma^2}$ at each antenna element. The inverse proportionality of the CRLB to the SNR is obvious. It is interesting to note that the CRLB depends on the direction of the incident wavefront itself. For $\alpha \to \pm\pi/2$, $\cos^2(\alpha) \to 0$ and the bound diverges to infinity. This is due to the visible aperture of the antenna array tends to zero for $\alpha \to \pm\pi/2$, which results in a vanishing angular resolution ability. Further, the bound is $\propto L^{-3}$. The reason for this strong dependency on the number of antenna elements is twofold. On the one hand, the available signal power increases with L. On the other hand, the antenna aperture increases as well as the number of antenna elements becomes higher.

3.3.3.1 Cramér–Rao Lower Bound for AOA Estimation using Linear Regression

We refer to Section 3.3.2.1 and calculate the CRLB for AOA estimation based on the argument $\rho_\ell = \arg(r_\ell)$ of the complex sample values $r_\ell = s\,\exp\left(j(\psi - \tilde{\ell}\,kd\sin(\alpha))\right) + n_\ell$, which we observe at antenna element $\ell = 0, \ldots, L-1$. Again, we use $\tilde{\ell} = \left(\ell - \frac{L-1}{2}\right)$ for notational convenience. The signal model in the angular domain is

$$\rho_\ell = \arg(r_\ell) = \psi - \tilde{\ell}\,kd\sin(\alpha) + v_\ell,$$

which we have already introduced in Section 3.3.2.1. We can approximate the phase noise term v_ℓ as real valued AWGN with zero mean and variance $E\{v_\ell^2\} = \sigma_v^2$. Thus, the likelihood function is

$$p(\rho|\alpha) = \left(2\pi\sigma_v^2\right)^{-\frac{L}{2}}\exp\left(-\frac{1}{2\sigma_v^2}\sum_{\ell=0}^{L-1}\left(\rho_\ell - \psi + \tilde{\ell}\,kd\sin(\alpha)\right)^2\right).$$

For the derivative of the log-likelihood function we get

$$\frac{d}{d\alpha}\ln p(\rho|\alpha) = -\frac{d}{d\alpha}\frac{1}{2\sigma_v^2}\sum_{\ell=0}^{L-1}\left(\rho_\ell - \psi + \tilde{\ell}\,kd\sin(\alpha)\right)^2$$

$$= -\frac{1}{\sigma_v^2}\sum_{\ell=0}^{L-1}\left(\rho_\ell - \psi + \tilde{\ell}\,kd\sin(\alpha)\right)\tilde{\ell}\,kd\cos(\alpha)$$

$$= -\frac{1}{\sigma_v^2}\sum_{\ell=0}^{L-1}v_\ell\,\tilde{\ell}\,kd\cos(\alpha).$$

Squaring and building the expectation value yields

$$E\left\{\left|\frac{d}{d\alpha}\ln p(\rho|\alpha)\right|^2\right\} = \frac{k^2d^2\cos^2(\alpha)}{\sigma_v^2}\sum_{\ell=0}^{L-1}\tilde{\ell}. \qquad (3.42)$$

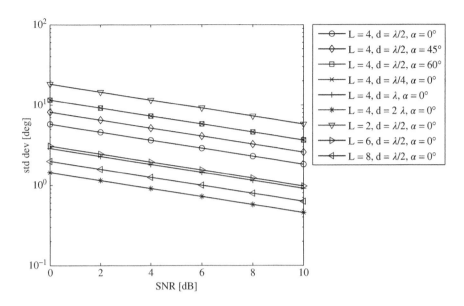

Figure 3.9 Comparison of CRLBs for AOA estimation performance

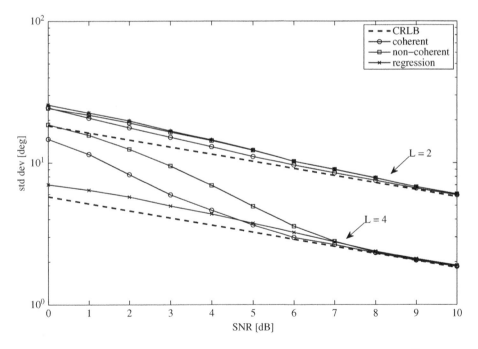

Figure 3.10 Comparison of AOA estimation performance, $d = \lambda/2, \alpha = 0°$

We use the approximation of the phase noise variance $\sigma_v^2 \approx \frac{\sigma^2}{2s^2} = \frac{1}{2\,\text{SNR}}$ and Equation (3.40). Inserting Equation (3.42) into Equation (3.12) finally results in the CRLB

$$\text{VAR}\,\{\hat{\alpha}(r)\} \geq \frac{6}{L(L^2 - 1)\,\text{SNR}\,k^2 d^2 \cos^2(\alpha)} \tag{3.43}$$

for angular estimation performance in radians. We observe exactly the same result as we got for the CRLB in the complex signal domain in Equation (3.41). This is because there is no angular information in the amplitude s at all. However, we should mention that Equation (3.43) assumes additive white Gaussian phase noise, which is a good approximation at high SNRs only.

Example 3.3.1 (AOA estimation performance) *Figure 3.9 shows the standard deviation* stddev $= \frac{180}{\pi} \sqrt{\text{VAR}\,\{\hat{\alpha}(r)\}}$ *in degrees for different antenna array parameters and wavefront incident angles versus the* SNR $= \frac{s^2}{\sigma^2}$ *at each antenna element. Besides the number of antenna elements, the product $kd \cos(\alpha)$ determines the visible antenna array aperture related to the wavelength λ. Thus, it is obvious that the parameter sets $kd \cos(\alpha) = \frac{2\pi}{\lambda} \frac{\lambda}{2} \cos(60°) = \frac{2\pi}{\lambda} \frac{\lambda}{4} \cos(0°)$ lead to equal CRLB results.*

In Figure 3.10, we compare numerical results for coherent, non-coherent, and linear regression based AOA estimation. We assume one signal propagation path, that is, there is one incident wavefront for which we have set $\alpha = 0°$. For $L = 4$, the linear regression method outperforms both coherent and non-coherent estimation at lower SNRs. In higher SNR regions, the performance of all estimation methods converges to the CRLB. Thus, performance differences at high SNRs are marginal. For two antenna elements, coherent estimation provides the best results. Performance differences to non-coherent and regression based AOA estimation as well as the divergence from the CRLB are smaller compared to $L = 4$.

4

Position Estimation

In this chapter, we address how to calculate the MT position, given a set of measurements according to Chapter 3 in a static scenario, using the position principles presented in Chapter 2. We assume that the MT is stationary during the position estimation process. Thus, the MT position is deterministic. In Chapter 5, we extend the static position estimation to the dynamic position tracking.

In this chapter, we consider ideal and erroneous measurements. With erroneous measurements, we mean measurements that are noisy or biased, for example, due to thermal noise or unknown MT clock offset for TOA measurements. However, we do not consider in this chapter measurement errors due to multipath and NLOS propagation. For that, we refer the reader to Chapter 7.

In the following section, we present the navigation equations for the positioning principles in Chapter 2. To allow a simple visualization, we restrict ourselves to a two-dimensional planar space. By default, we assume Cartesian coordinates. Where appropriate, we will extend our considerations to three-dimensional space.

The BSs are located at

$$\mathbf{x}_i = \begin{pmatrix} x_i & y_i \end{pmatrix}^{\mathrm{T}}, \qquad\qquad i = 1, \dots, N$$

and the MT is located at

$$\mathbf{x} = \begin{pmatrix} x & y \end{pmatrix}^{\mathrm{T}}.$$

The distance between i-th BS and the MT is given by

$$d_i = d_i(\mathbf{x}) = ||\mathbf{x} - \mathbf{x}_i||_2 = \sqrt{(x - x_i)^2 + (y - y_i)^2}.$$

4.1 Triangulation

In triangulation, the MT and two BSs form a triangle, where the length and orientation of the base line d_{12} between the two BSs is known and the AOAs $\varphi_1 = \angle BS_2 BS_1 MT$ and $\varphi_2 = \angle MTBS_2 BS_1$ are measured in uplink transmissions from the MT to the BSs (see Figure 4.1).

Positioning in Wireless Communications Systems, First Edition. Stephan Sand, Armin Dammann and Christian Mensing.
© 2014 John Wiley & Sons, Ltd. Published 2014 by John Wiley & Sons, Ltd.

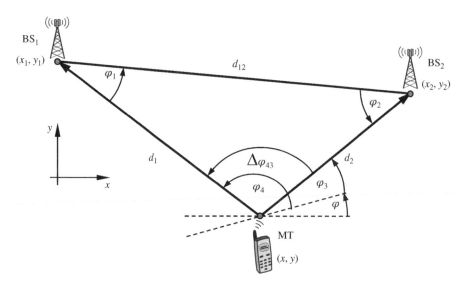

Figure 4.1 Triangulation with AOA measurements

Alternative, the angles φ_3, φ_4, and φ can be measured at the MT. φ measures the orientation of the MT with respect to its antenna orientation and the dashed line parallel to the x-axis containing the MT.

4.1.1 Triangulation with Ideal Measurements

In this section, we derive and discuss the navigation equations for ideal angular measurements. We assume that the orientation of antennas is known at the BS as their positions are fixed. In Section 2.2, we introduced the principle of AOA based positioning distinguishing two cases, that is, downlink AOA with known or unknown MT orientation.

4.1.1.1 Downlink AOA with Known MT Orientation or Uplink AOA

When considering downlink BS to MT AOA measurements, the orientation of the MT needs to be known. This may be achieved with additional sensors, for example, a magnetometer. Equivalently, we consider uplink MT to BS AOA measurements. The navigation equations relating the N AOA measurements and the MT and BS coordinates are then according to Equation (2.16)

$$y - y_i = \tan(\varphi_i)(x - x_i), \qquad i = 1, \ldots N \qquad (4.1)$$

for the position (x, y) of the MT. We infer from Equation (4.1) that we have N linearly independent equations for the two unknowns x and y. As Equation (4.1) is a linear equation system, two AOA measurements φ_1 and φ_2 would suffice to uniquely solve for the MT position (x, y). Each measurement equation in Equation (4.1) defines a straight line in two-dimensional space (see Figures 2.7 and 4.1). If there is no error in the measurements, all lines intersect in a single point, which is the MT position (x, y).

Analytical Solution to Triangulation Equations

To solve Equation (4.1), the BSs measure the AOAs φ_1 and φ_2 in Figure 4.1 (see Section 3.3). Further, the coordinates (x_1, y_1) and (x_2, y_2) of the two BSs need to be known. Then, we can relate the MT coordinates with the measured angles and the BS coordinates as

$$\tan \varphi_1 = \frac{y - y_1}{x - x_1} \qquad \text{and} \qquad \tan \varphi_2 = \frac{y - y_2}{x - x_2}.$$

Solving this equation for y, we obtain

$$y = (\tan \varphi_1)x + y_1 - \tan \varphi_1 x_1 = a_1 x + b_1,$$

$$y = (\tan \varphi_2)x + y_2 - \tan \varphi_2 x_2 = a_2 x + b_2. \qquad (4.2)$$

These two equations represent two lines, whose intersection point (x, y) is the MT position. Solving Equation (4.2) for the constants b_1 and b_2, and using vector-matrix notation, we obtain

$$\begin{pmatrix} -a_1 & 1 \\ -a_2 & 1 \end{pmatrix} \begin{pmatrix} x \\ y \end{pmatrix} = \begin{pmatrix} b_1 \\ b_2 \end{pmatrix}. \qquad (4.3)$$

To solve the linear equation system (4.3), we simply invert the 2×2 matrix

$$\begin{pmatrix} x \\ y \end{pmatrix} = \begin{pmatrix} -a_1 & 1 \\ -a_2 & 1 \end{pmatrix}^{-1} \begin{pmatrix} b_1 \\ b_2 \end{pmatrix} = \frac{1}{a_2 - a_1} \begin{pmatrix} 1 & a_2 \\ -1 & -a_1 \end{pmatrix} \begin{pmatrix} b_1 \\ b_2 \end{pmatrix}$$

$$= \frac{1}{\tan \varphi_2 - \tan \varphi_1} \begin{pmatrix} y_1 - \tan \varphi_1 x_1 + \tan \varphi_2 \left(y_2 - \tan \varphi_2 x_2 \right) \\ -y_1 + \tan \varphi_1 x_1 - \tan \varphi_1 (y_2 - \tan \varphi_2 x_2) \end{pmatrix}. \qquad (4.4)$$

Note that Equation (4.4) only holds if $a_1 = \tan \varphi_1 \neq a_2 = \tan \varphi_2$, that is, the two AOAs are not measured from exactly the same or opposite direction.

4.1.1.2   Downlink AOA with Unknown MT Orientation

In general, the orientation φ of the MT is unknown when using downlink AOA measurements (see Figure 4.1). Thus, it becomes an additional parameter that needs to be estimated to obtain the MT position (x, y). Therefore, Equation (4.1) becomes

$$y - y_i = \tan(\varphi_i + \varphi)(x - x_i), \qquad\qquad i = 1, \dots N. \qquad (4.5)$$

As explained in Section 2.2.1, Equation (4.5) is nonlinear in φ. When φ is unknown and we have only two AOA measurements φ_1 and φ_2, the location of the MT is limited to an arc of a circle inscribing a triangle defined with baseline $d = \sqrt{(x_1 - x_2)^2 + (y_1 - y_2)^2}$ and angle $\varphi_2 - \varphi_1$ opposite the baseline d. Thus, at least three AOA measurements are needed to estimate the MT position uniquely in contrast to Equation (4.1). Alternative, we can rearrange Equation (4.5) so it is linear with respect to the angles φ_i and φ, but nonlinear with respect to the coordinates x and y:

$$\arctan \left(\frac{y - y_i}{x - x_i} \right) = \varphi_i + \varphi, \qquad\qquad i = 1, \dots N. \qquad (4.6)$$

Thus, the unknown MT orientation in Equation (4.6) is similar to the unknown clock bias b between the MT and BSs for TOA measurements (see Section 2.1.1.2). For TOA measurements, we computed the difference between a reference TOA and the other TOA measurements to remove the unknown clock bias b. This resulted in TDOA based positioning (see Section 2.1.2). Thus, we can subtract Equation (4.6) for $i = 1$ from Equation (4.6) for $i = 2, \ldots, N$ to obtain

$$\Delta\varphi_{i1} = \arctan\left(\frac{y - y_i}{x - x_i}\right) - \arctan\left(\frac{y - y_1}{x - x_1}\right), \qquad i = 2, \ldots N. \qquad (4.7)$$

Here, $\Delta\varphi_{i1} = \varphi_i - \varphi_1$ denotes the *AOA difference (AOAD)* measurements.

In the following, we present an alternative approach for AOA positioning with unknown MT orientation φ. For that, we use the law of cosines, that is,

$$c^2 = a^2 + b^2 - 2ab\cos\Delta\varphi,$$

to obtain the set of nonlinear equations

$$d_{1i}^2 = d_1^2 + d_i^2 - 2d_1 d_i \cos(\varphi_i - \varphi_1)$$

$$(x_1 - x_i)^2 + (y_1 - y_i)^2 = (x - x_1)^2 + (y - y_1)^2 + (x - x_i)^2 + (y - y_i)^2$$

$$- 2\sqrt{(x - x_1)^2 + (y - y_1)^2}\sqrt{(x - x_i)^2 + (y - y_i)^2} \cdot \underbrace{\cos(\varphi_i - \varphi_1)}_{=:\Delta\varphi_{i1}},$$

$$i = 2, \ldots, N. \qquad (4.8)$$

The cosines in Equation (4.8) are calculated for AOAD measurements $\Delta\varphi_{i1} = \varphi_i - \varphi_1$ from the AOA measurements φ_i, $i = 2, \ldots, N$ and the arbitrarily chosen AOA reference measurement φ_1. In Figure 4.1, we used φ_3 as reference measurement. Thus, the orientation φ of the MT is removed compared to Equation (4.5) and only $N - 1$ equations remain. This is similar to TDOA based positioning, when the unknown clock bias b between the MT and the BS for TOA measurements is removed by computing the difference between a reference TOA and the other TOA measurements (see Section 2.1.2). Similar to Equations (4.5) and (4.6), we can solve Equation (4.8) for $\Delta\varphi_{i1}$

$$\Delta\varphi_{i1} =$$

$$\arccos\left(\frac{(x - x_1)^2 + (y - y_1)^2 + (x - x_i)^2 + (y - y_i)^2 - (x_1 - x_i)^2 - (y_1 - y_i)^2}{2\sqrt{(x - x_1)^2 + (y - y_1)^2}\sqrt{(x - x_i)^2 + (y - y_i)^2}}\right)$$

$$i = 2, \ldots, N. \qquad (4.9)$$

The benefits of Equations (4.6), (4.7), and (4.9) will be explained in Section 4.1.2.

We can obtain a polynomial equation system with respect to the unknowns x and y from Equation (4.8) by solving for the square root terms and squaring each equation, that is,

$$4\left((x - x_1)^2 + (y - y_1)^2\right)\left((x - x_i)^2 + (y - y_i)^2\right) \cdot \cos^2(\Delta\varphi_{i1}) =$$

$$\left((x - x_1)^2 + (y - y_1)^2 + (x - x_i)^2 + (y - y_i)^2 - (x_1 - x_i)^2 - (y_1 - y_i)^2\right)^2,$$

$$i = 2, \ldots, N. \qquad (4.10)$$

Comparing Equation (4.10) with Equations (4.7) or (4.9), we note that the first equation system is polynomial in x and y whereas the later ones are transcendental. Thus, the navigation equation (4.10) promises a lower complexity and possibly a more numerically stable solution than Equation (4.7). Nevertheless, Equations (4.7) and (4.9) are beneficial for erroneous measurements (see Section 4.1.2).

Given three independent AOA measurements from three different BSs, we obtain two independent polynomial equations with maximum degree of four for both x and y according to Equation (4.10). As two circles intersect at most at two places the resulting four circles intersect at most at 12 places.

Example 4.1.1 (AOAD solution geometry) *Here, BS_1 is located at $(x, y) = (0\,\text{m}, 0\,\text{m})$, BS_2 at $(100\,\text{m}, 0\,\text{m})$, and BS_3 at $(0\,\text{m}, 100\,\text{m})$. The MT measured the AOADs $\Delta\varphi_{21} = 150°$ and $\Delta\varphi_{31} = -75°$. The solution geometry defined by Equation (4.10) is plotted in Figure 4.2. As explained, the two polynomial equations with a maximum degree of 4 in Equation (4.10) result in two pairs of circles (one small and one large pair) that intersect at 12 points. The intersection points $1, 2, 3, 4, BS_2$, and BS_3 have a multiplicity of one and the intersection point BS_1 a multiplicity of six. Due to the squaring to obtain Equation (4.10), any of the intersection points between the pair of small and large circles, that is, $1, 2, 3, 4$, and BS_1, could be a possible position of the MT. However, assuming a non-zero distance to BS_1 and given the AOAD measurements $\Delta\varphi_{21} = 150°$ and $\Delta\varphi_{31} = -75°$, only point 4 is the valid position solution.*

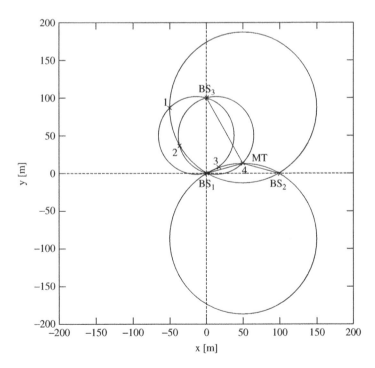

Figure 4.2   AOAD measurements from three BSs and solution geometry according to Equation (4.10)

Numerical Solution to Triangulation with Unknown MT Orientation
In this case, we have to solve one of the nonlinear Equations (4.5)–(4.10). To the best of
our knowledge, only the Equations (4.8) and (4.10) might have rather complex closed form
solutions. Nevertheless, we will present here two iterative algorithms, the Newton–Raphson
and Gauss–Newton algorithms (Kay 1993; Press et al. 1992), to solve the nonlinear navigation
equations of Section 4.1 adapted to the *nonlinear least squares* problem. Apart from these two
algorithms, other algorithms such as steepest descent or Levenberg–Marquardt exist to solve
the nonlinear navigation equations (see Sections 4.2.2.3 and 4.3.2.1).

Newton–Raphson algorithm.   Given the previous equations for unknown MT orientation, we
could try to directly solve the nonlinear equations, that is,

$$f(\theta) = \left(f_1(\theta) \ \cdots \ f_N(\theta) \right)^{\mathrm{T}}, \tag{4.11}$$

where $f_i(\theta) = 0$ corresponds to the i-th element in Equations (4.5)–(4.10), for example,
$f_i(\theta) = f_i((x, \varphi)^{\mathrm{T}}) = y - y_i - \tan(\varphi_i + \varphi)(x - x_i) = 0$ for Equation (4.5). Note that for
Equations (4.7) or (4.8)–(4.10) we have, instead of N, only $N - 1$ equations. Nevertheless,
the following derivations and equations also hold for Equations (4.7)–(4.10) assuming that
we have reduced the dimensions of vectors and matrices accordingly. The vector $\theta \in \mathbb{R}^M$
contains the M unknowns, for example, x, y, and φ for Equation (4.5). Instead of directly
computing the solution to Equation (4.11), it is numerically more stable to minimize the
quadratic cost function (Press et al. 1992)

$$C(\theta) = \frac{1}{2} f^{\mathrm{T}}(\theta) f(\theta) = \frac{1}{2} \sum_{i=0}^{N} f_i^2(\theta). \tag{4.12}$$

Note that $C(\theta)$ defines the *nonlinear least squares* problem (Kay 1993). Next, we compute the
gradient of $C(\theta)$ and set it equal to zero to minimize Equation (4.12):

$$g(\theta) = \nabla_\theta C(\theta) = \sum_{i=0}^{N} \nabla_\theta f_i(\theta) f_i(\theta) = \underbrace{(\nabla_\theta \otimes f^{\mathrm{T}}(\theta)) f(\theta)}_{=: \Phi^{\mathrm{T}}(\theta)} = \Phi^{\mathrm{T}}(\theta) f(\theta) \stackrel{!}{=} \mathbf{0}_M, \tag{4.13}$$

where $\nabla_\theta = \left(\frac{\partial}{\partial \theta_1}, \ \cdots \ \frac{\partial}{\partial \theta_M} \right)^{\mathrm{T}}$, $\otimes$ denotes the Kronecker product, and $\Phi(\theta)$ the $N \times M$
Jacobian matrix. Note, Equation (4.13) is a system of M simultaneous nonlinear equations
with respect to θ. To solve these equations, we apply the Newton–Raphson method (Kay
1993) to the function $g(\theta)$ in Equation (4.13). Then, the k-th iteration is given by

$$\theta^{(k+1)} = \theta^{(k)} - \left((\Gamma(\theta))^{-1} g(\theta) \right) \Big|_{\theta = \theta^{(k)}}, \tag{4.14}$$

with the $M \times M$ Jacobian matrix

$$\Gamma(\theta) = \nabla_\theta^{\mathrm{T}} \otimes g(\theta) = \nabla_\theta^{\mathrm{T}} \otimes ((\nabla_\theta \otimes f^{\mathrm{T}}(\theta)) f(\theta))$$

$$= \Phi^{\mathrm{T}}(\theta) \Phi(\theta) + \sum_{i=1}^{N} f_i(\theta) H_{f_i(\theta)} \tag{4.15}$$

and the $M \times M$ Hessian matrix

$$
\boldsymbol{H}_{f_i(\boldsymbol{\theta})} = (\nabla_{\boldsymbol{\theta}} \otimes \nabla_{\boldsymbol{\theta}}^{\mathrm{T}}) f_i(\boldsymbol{\theta}) =
\begin{pmatrix}
\frac{\partial^2 f_i(\boldsymbol{\theta})}{\partial \theta_1^2} & \frac{\partial^2 f_i(\boldsymbol{\theta})}{\partial \theta_1 \partial \theta_2} & \cdots & \frac{\partial^2 f_i(\boldsymbol{\theta})}{\partial \theta_1 \partial \theta_M} \\
\frac{\partial^2 f_i(\boldsymbol{\theta})}{\partial \theta_2 \partial \theta_1} & \frac{\partial^2 f_i(\boldsymbol{\theta})}{\partial \theta_2^2} & \cdots & \frac{\partial^2 f_i(\boldsymbol{\theta})}{\partial \theta_2 \partial \theta_M} \\
\vdots & & \ddots & \vdots \\
\frac{\partial^2 f_i(\boldsymbol{\theta})}{\partial \theta_M \partial \theta_1} & \cdots & \frac{\partial^2 f_i(\boldsymbol{\theta})}{\partial \theta_M \partial \theta_{M-1}} & \frac{\partial^2 f_i(\boldsymbol{\theta})}{\partial \theta_M^2}
\end{pmatrix}.
\tag{4.16}
$$

Example 4.1.2 (Newton–Raphson iteration for triangulation with AOAD) *In the following, we apply the Newton–Raphson iterations (see Equations 4.14–4.16) to the AOAD Equation (4.10) and the parameters of Example 4.1.1. Thus, the parameter vector only contains the x- and y-coordinates of the MT, that is, $\boldsymbol{\theta} = (x, y)^{\mathrm{T}}$. The function $f(\boldsymbol{\theta})$ then becomes*

$$
f(\boldsymbol{\theta}) = 4 \begin{pmatrix} \left(x^2 + y^2 - xx_2\right)^2 - \cos^2\left(\Delta\varphi_{21}\right)\left(x^2 + y^2\right)\left(x^2 + y^2 - 2xx_2 + x_2^2\right) \\ \left(x^2 + y^2 - yy_3\right)^2 - \cos^2\left(\Delta\varphi_{31}\right)\left(x^2 + y^2\right)\left(x^2 + y^2 - 2yy_3 + y_3^2\right) \end{pmatrix},
$$

with $x_2 = y_3 = 100\,\mathrm{m}$, $\Delta\varphi_{21} = 150°$, and $\Delta\varphi_{31} = -75°$. Thus, given $f(\boldsymbol{\theta})$ in the previous equation, we can compute a Newton–Raphson iteration by calculating

1. $\boldsymbol{g}(\boldsymbol{\theta})$ in Equation (4.13),
2. $\boldsymbol{\Gamma}(\boldsymbol{\theta})$ in Equation (4.15), and
3. finally, evaluating Equation (4.14) at $\boldsymbol{\theta} = \boldsymbol{\theta}^{(k)}$ to obtain the estimate $\boldsymbol{\theta}^{(k+1)}$.

We start this algorithm with an initial value of $\boldsymbol{\theta}(0)$. Then, we repeat the three steps until a new update changes less than a predefined tolerance ϵ_{TOL}, that is,

$$
\left\| \boldsymbol{\theta}^{(k+1)} - \boldsymbol{\theta}^{(k)} \right\|_2 < \epsilon_{\mathrm{TOL}}.
$$

Figure 4.3 shows the scenario including the Newton–Raphson iterations (+ markers). We initialize the algorithm with the starting point $(100\,\mathrm{m}, 100\,\mathrm{m})$. The algorithm converges to the correct MT position at $(50\,\mathrm{m}, 13.4\,\mathrm{m})$ after 11 iterations. Note that the algorithm is very sensitive to the starting point. For instance, it converges to the point $(16.14\,\mathrm{m}, 7.5\,\mathrm{m})$ if we chose the mean position of the measured BSs, that is, $\hat{\boldsymbol{x}}^{(0)} = 1/N \sum_{i=1}^{N} \boldsymbol{x}_i = (33.3\,\mathrm{m}, 33.3\,\mathrm{m})^{\mathrm{T}}$. This point is a valid solution to Equation (4.10). However, if we check the estimated solution with the measured AOADs, we see that $\Delta\varphi_{31} = 105°$ instead of the correct 75°. Thus, the position estimation could chose another starting point until the position solution yields the correct AOADs.

Gauss–Newton Algorithm. Instead of directly working with the nonlinear function $f(\boldsymbol{\theta})$, we now linearize Equation (4.11) at $\boldsymbol{\theta}^{(k)}$:

$$
f(\boldsymbol{\theta}) \approx \tilde{f}(\boldsymbol{\theta}) = f\left(\boldsymbol{\theta}^{(k)}\right) + \boldsymbol{\Phi}(\boldsymbol{\theta})|_{\boldsymbol{\theta}=\boldsymbol{\theta}^{(k)}} \left(\boldsymbol{\theta} - \boldsymbol{\theta}^{(k)}\right).
\tag{4.17}
$$

Thus the $N \times M$ Jacobian matrix of the linearized function $\tilde{f}(\boldsymbol{\theta})$ is

$$
\tilde{\boldsymbol{\Phi}}(\boldsymbol{\theta}) = \nabla_{\boldsymbol{\theta}}^{\mathrm{T}} \otimes \tilde{f}(\boldsymbol{\theta}) = \boldsymbol{\Phi}(\boldsymbol{\theta})|_{\boldsymbol{\theta}=\boldsymbol{\theta}^{(k)}}.
\tag{4.18}
$$

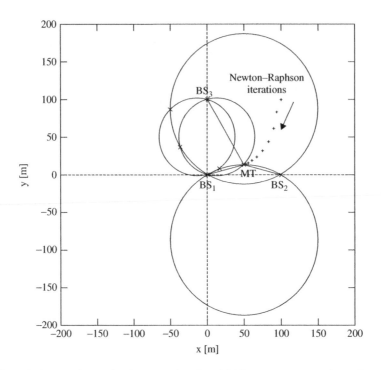

Figure 4.3 Solution to the navigation equation for AOAD measurements from three BSs with Newton–Raphson method: Pluses indicate the Newton–Raphson iterations starting at (100 m, 100 m). Convergence is achieved after 11 iterations

Consequently, the function

$$\tilde{g}(\boldsymbol{\theta}) = \boldsymbol{\Phi}^T\left(\boldsymbol{\theta}\right)\big|_{\boldsymbol{\theta}=\boldsymbol{\theta}^{(k)}}\left(\boldsymbol{f}\left(\boldsymbol{\theta}^{(k)}\right) + \boldsymbol{\Phi}(\boldsymbol{\theta})\big|_{\boldsymbol{\theta}=\boldsymbol{\theta}^{(k)}}\left(\boldsymbol{\theta} - \boldsymbol{\theta}^{(k)}\right)\right) \tag{4.19}$$

is now linear in $\boldsymbol{\theta}$. Hence, setting $\tilde{g}(\boldsymbol{\theta}) \overset{!}{=} 0$ and solving for $\boldsymbol{\theta}$ yields the Gauss–Newton iteration

$$\boldsymbol{\theta}^{(k+1)} = \boldsymbol{\theta}^{(k)} - \left(\left(\boldsymbol{\Phi}^T\left(\boldsymbol{\theta}\right)\boldsymbol{\Phi}(\boldsymbol{\theta})\right)^{-1}\boldsymbol{\Phi}^T\left(\boldsymbol{\theta}\right)\right)\Big|_{\boldsymbol{\theta}=\boldsymbol{\theta}^{(k)}}\boldsymbol{f}\left(\boldsymbol{\theta}^{(k)}\right). \tag{4.20}$$

4.1.1.3 Extension to Three Dimensions

In the remainder of this section, we extend these considerations to three-dimensional navigation (see Section 2.2.2). Figure 2.9 depicts the three-dimensional setup. In three-dimensional space, we can measure, in addition to the two azimuth angles φ_1 and φ_2, two elevation angles ϑ_1 and ϑ_2. The relations between the x- and y-coordinates in Equation (4.5) still hold in three-dimensional space. Thus, we can obtain from Section 2.2.2 and Figure 2.9 the following relations for the azimuth and elevation angles:

$$\begin{aligned}
\left(y - y_i\right) &= \tan\left(\varphi_i + \varphi\right)\left(x - x_i\right), \\
\cos\left(\varphi_i + \varphi\right)\left(z - z_i\right) &= \tan\left(\vartheta_i + \vartheta\right)\left(x - x_i\right), \qquad i = 1, \dots, N,
\end{aligned} \tag{4.21}$$

or with $d_{i,2D} = \sqrt{(x - x_i)^2 + (y - y_i)^2}$ for the second line in Equation (4.21)

$$(z - z_i) = \tan(\vartheta_i + \vartheta) d_{i,2D}, \qquad i = 1, \ldots, N. \qquad (4.22)$$

In Equation (4.21), we have the five unknowns x, y, z, φ, and ϑ. Thus, at least five AOA measurements are needed, for example, three azimuth and two elevation AOAs. Similar to Equations (4.5), (4.6), and (4.7), we formulate the modified navigation equations for Equation (4.21) as

$$\arctan\left(\frac{y - y_i}{x - x_i}\right) = \varphi_i + \varphi,$$

$$\arctan\left(\frac{z - z_i}{d_{i,2D}}\right) = \vartheta_i + \vartheta, \qquad i = 1, \ldots, N. \qquad (4.23)$$

and as

$$\arctan\left(\frac{y - y_i}{x - x_i}\right) - \arctan\left(\frac{y - y_1}{x - x_1}\right) = \Delta\varphi_{i1}$$

$$\arctan\left(\frac{z - z_i}{d_{i,2D}}\right) - \arctan\left(\frac{z - z_1}{d_{1,2D}}\right) = \Delta\vartheta_{i1} \qquad i = 2, \ldots, N. \qquad (4.24)$$

Besides Equations (4.23) and (4.24), we can alternatively compute the z-coordinate of the MT position according to Figure 4.4 and the law of cosines as

$$
\begin{aligned}
a_i^2 &= (x - x_i)^2 + (y - y_i)^2 + (z - z_i)^2, \\
b_i^2 &= (x - x_1)^2 + (y - y_1)^2 + (z - z_1)^2, \\
c_i^2 &= \left(\sqrt{(x - x_i)^2 + (y - y_i)^2} - \sqrt{(x - x_1)^2 + (y - y_1)^2}\right)^2 + (z_i - z_1)^2, \\
&= a_i^2 + b_i^2 - 2a_i b_i \cos(\Delta\vartheta_{i1}), \qquad i = 2, \ldots, N,
\end{aligned}
\qquad (4.25)
$$

or

$$\Delta\vartheta_{i1} = \arccos\left(\frac{a_i^2 + b_i^2 - c_i^2}{2a_i b_i}\right), \qquad i = 2, \ldots, N, \qquad (4.26)$$

where $\Delta\vartheta_{i1} = \vartheta_i - \vartheta_1$ is the elevation AOAD. Thus, we can obtain the z-coordinate of the MT position with one AOAD measurement from Equation (4.25), which is an algebraic equation in z. By squaring the last line of Equation (4.25), we obtain a fourth order polynomial equation in z, that is, a quartic equation, which can be solved analytically.

As the relations between the azimuth angles and the x- and y-coordinates are independent of the z-coordinate, we can compute the position of the MT in two steps:

1. We compute the Cartesian coordinates of the MT in the xy-plane according to Equation (4.1) for known φ or according to Equations (4.5)–(4.10) for unknown φ.
2. We compute the z-coordinate of the MT according to one of the Equations (4.21)–(4.26) given the estimates for x, y, and possibly φ, and at least one elevation AOA ϑ_i for known ϑ, or two elevation AOAs for unknown ϑ.

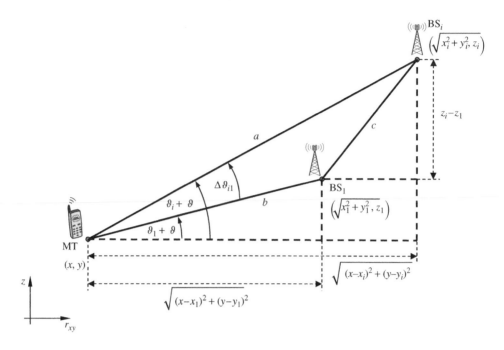

Figure 4.4 Step 2 of 3D AOA estimation with AOAD $\Delta\vartheta_{i1}$

Note that for computing the three-dimensional position of the MT, we only need to measure three angles, for example, the two azimuth angles φ_1 and φ_2 as well as the elevation angle ϑ_1 if the MT orientation is known. From Figure 2.9, we may geometrically interpret this as follows: When computing the x- and y-coordinates of the MT first, the three-dimensional position solution to the MT lies on the intersection of two planes. These planes are perpendicular to the xy-plane and contain the points (x_1, y_1) and (x, y) or (x_2, y_2) and (x, y). The intersection of the two planes forms a line parallel to the z-coordinate, on which the MT lies. Given one elevation angle as third measurement, we uniquely determine the MT position (x, y, z). However, we need to measure at least five angles if the MT orientation is unknown, that is, $\varphi_1, \varphi_2, \varphi_3, \vartheta_1$, and ϑ_2.

Analytical Solution to the Triangulation Equations with Known MT Orientation
The previous calculations show that it is sufficient to just have two azimuth angle measurements for planar navigation if the MT orientation is known or uplink AOAs are measured at the BSs. In the case of three-dimensional navigation, three angular measurements, for example, two azimuth angles and one elevation angle, are necessary to compute the MT position. Thus, the z-coordinate of the MT position is calculated from Equation (4.21) as

$$z = z_1 + \tan(\vartheta_1)d_{1,2D},$$

where $d_{1,2D} = \sqrt{(x - x_1)^2 + (y - y_1)^2}$ can be calculated given the solution from Equation (4.4).

Numerical Solution to the Triangulation Equations with Unknown MT Orientation
In this case, we have to solve one of the nonlinear Equations (4.21)–(4.26). To the best of our knowledge, only Equation (4.25) might have a rather complex closed form solutions.

In Section 4.1.1.2, we presented the Newton–Raphson and Gauss–Newton algorithm for the two-dimensional Equations (4.5)–(4.10). The extension to the three-dimensional Equations (4.21)–(4.26) is straightforward. We simply have to modify $f(\theta)$ in Equation (4.11) and then, subsequently in Equations (4.12)–(4.20).

4.1.2 Triangulation with Erroneous Measurements

In contrast to the previous section, any practical system cannot provide ideal measurements. The measurements are erroneous due to various effects, such as thermal noise in the electric circuits, limited angular resolution of the measurement equipment, uncertainties in the BS positions, and so on. A common assumption is that the AOA measurement errors are disturbed by AWGN (see Section 3.3.2, Figueiras and Frattasi 2010; Zekavat and Buehrer 2012):

$$\hat{\varphi}_i = \varphi_i + \varphi_{\epsilon_i}, \qquad \varphi_{\epsilon_i} \sim \mathcal{N}\left(0, \sigma^2_{\varphi_{\epsilon_i}}\right), \qquad i = 1, \dots, N \qquad (4.27)$$

and

$$\hat{\vartheta}_i = \vartheta_i + \vartheta_{\epsilon_i}, \qquad \vartheta_{\epsilon_i} \sim \mathcal{N}\left(0, \sigma^2_{\vartheta_{\epsilon_i}}\right), \qquad i = 1, \dots, N. \qquad (4.28)$$

Hence, the navigation equations for triangulation with erroneous measurements can be simply obtained from the ones for ideal measurements through replacing φ_i with $\hat{\varphi}_i$ and ϑ_i with $\hat{\vartheta}_i$ in Equations (4.1), (4.5)–(4.10) and (4.21)–(4.26). Note that the AWGN on the AOAs $\hat{\varphi}_i$ and $\hat{\vartheta}_i$ in Equations (4.27) and (4.28) is transformed through the trigonometric functions cos and tan to a multiplicative, non-Gaussian noise in the triangulation equations, that is, in Equations (4.1), (4.5), (4.8), (4.10), (4.21), (4.22), and (4.25). This will complicate the derivation of position solution algorithms, especially, the ML algorithms. Thus, the triangulation equations defined by Equations (4.6), (4.7), (4.9), (4.23), (4.24), and (4.26) are more suitable for noisy AOA measurements.

Figure 4.5 depicts a possible measurement scenario for two-dimensional navigation with noisy measurements. Let us assume that N BS have measured the AOA of the MT received signal, that is, $\hat{\varphi}_i$ and $\hat{\vartheta}_i$ $(i = 1, \dots, N)$ according to Equation (4.27) and (4.28).

4.1.2.1 Linear Least Squares Solution to Erroneous AOA Measurements with Known MT Orientation

Although Equations (4.1) or (4.21) only hold exactly for ideal measurements, a simple way of solving the triangulation equations with erroneous measurements is to apply the least squares method to Equations (4.1) or (4.21). With Equation (4.1), we can write the linear equation system according to Equation (4.3) as

$$\underbrace{\begin{pmatrix} -\tan\hat{\varphi}_1 & 1 \\ -\tan\hat{\varphi}_2 & 1 \\ \vdots & \vdots \\ -\tan\hat{\varphi}_N & 1 \end{pmatrix}}_{=:A} \underbrace{\begin{pmatrix} x \\ y \end{pmatrix}}_{=:\theta} = \underbrace{\begin{pmatrix} y_1 - \tan\hat{\varphi}_1 x_1 \\ y_2 - \tan\hat{\varphi}_2 x_2 \\ \vdots \\ y_N - \tan\hat{\varphi}_N x_N \end{pmatrix}}_{=:B}$$

$$A\theta = B. \qquad (4.29)$$

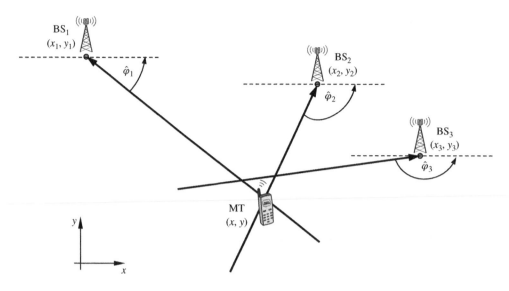

Figure 4.5   Triangulation of MT with noisy measurements in two-dimensional Cartesian space

Since the matrix A has dimensions $N \times 2$, Equation (4.29) has no unique solution. Neverthe-less, we can compute the *linear least squares solution* to Equation (4.29) according to (Haykin 2002; Kay 1993) as

$$\theta = \left(A^{\mathrm{T}}A\right)^{-1}A^{\mathrm{T}}B. \tag{4.30}$$

Unfortunately, due to the complicated influence of the noise in Equation (4.29), we cannot determine any statistical properties of the least squares estimator in Equation (4.30).

When considering three-dimensional position estimation, we obtain an approximate set of linear equations with Equation (4.21)

$$\underbrace{\begin{pmatrix} -\tan\hat{\varphi}_1 & 1 & 0 \\ -\tan\hat{\varphi}_2 & 1 & 0 \\ \vdots & \vdots & \vdots \\ -\tan\hat{\varphi}_N & 1 & 0 \\ -\tan\hat{\vartheta}_1\sqrt{1+\tan^2\hat{\varphi}_1} & 0 & 1 \\ -\tan\hat{\vartheta}_2\sqrt{1+\tan^2\hat{\varphi}_2} & 0 & 1 \\ \vdots & \vdots & \vdots \\ -\tan\hat{\vartheta}_N\sqrt{1+\tan^2\hat{\varphi}_N} & 0 & 1 \end{pmatrix}}_{=:A} \underbrace{\begin{pmatrix} x \\ y \\ z \end{pmatrix}}_{=:\theta} = \underbrace{\begin{pmatrix} y_1 - \tan\hat{\varphi}_1 x_1 \\ y_2 - \tan\hat{\varphi}_2 x_2 \\ \vdots \\ y_N - \tan\hat{\varphi}_N x_N \\ z_1 - \tan\hat{\vartheta}_1\sqrt{1+\tan^2\hat{\varphi}_1} x_1 \\ z_2 - \tan\hat{\vartheta}_2\sqrt{1+\tan^2\hat{\varphi}_2} x_2 \\ \vdots \\ z_N - \tan\hat{\vartheta}_N\sqrt{1+\tan^2\hat{\varphi}_N} x_N \end{pmatrix}}_{=:B}$$

$$A\theta = B.$$

Given this equation, we can compute the least squares estimate of θ with Equation (4.30).

4.1.2.2 Maximum Likelihood Solution to Erroneous Downlink AOA Measurements with Unknown MT Orientation

In this section, we develop the ML position estimator for erroneous AOA measurements in the downlink with unknown MT orientation. Similar to Section 4.1.1.2 we present the iterative Newton–Raphson and Gauss–Newton algorithms to solve the *weighted nonlinear least squares* problem when the AOA measurements are corrupted by AWGN.

Given the navigation equation in Equation (4.6), we define the vector function

$$f(\theta) = \left(f_1(\theta) \; \cdots \; f_N(\theta) \right)^{\mathrm{T}}, \qquad (4.31)$$

where $f_i(\theta)$ is given by

$$f_i(\theta) = \arctan\left(\frac{y - y_i}{x - x_i} \right) - \varphi, \qquad\qquad i = 1, \ldots, N.$$

The measurement vector is then

$$\hat{\varphi} = f(\theta) + \varphi_\epsilon = \left(f_1(\theta) + \varphi_{\epsilon_1} \; \cdots \; f_N(\theta) + \varphi_{\epsilon_N} \right)^{\mathrm{T}},$$

with

$$\varphi_\epsilon \sim \mathcal{N}\left(0_N, \Sigma_{\varphi_\epsilon} \right) \qquad \text{and} \qquad \Sigma_{\varphi_\epsilon} = \mathrm{E}\left\{ \varphi_\epsilon \varphi_\epsilon^{\mathrm{T}} \right\}.$$

If the AOA measurements are uncorrelated, $\Sigma_{\varphi_\epsilon} \in \mathbb{R}^{N \times N}$ becomes a diagonal matrix. For the random vector $\hat{\varphi}$, the log-likelihood function given $f(\theta)$ is

$$\ln \mathrm{p}(\hat{\varphi}|f(\theta)) = -\frac{N}{2} \ln(2\pi) - \frac{1}{2} \ln \det \Sigma_{\varphi_\epsilon} - \frac{1}{2} \left((\hat{\varphi} - f(\theta))^{\mathrm{T}} \Sigma_{\varphi_\epsilon}^{-1} (\hat{\varphi} - f(\theta)) \right).$$

Only the last term in the previous equation is a function of θ. Thus, we define the ML cost function as

$$L(\theta) = -\frac{1}{2} (\hat{\varphi} - f(\theta))^{\mathrm{T}} \Sigma_{\varphi_\epsilon}^{-1} (\hat{\varphi} - f(\theta)). \qquad (4.32)$$

This cost function defines the *weighted nonlinear least squares* problem. To find the ML solution for Equation (4.32), we compute the gradient of $L(\theta)$ and equate it to the zero vector:

$$g(\theta) = \nabla_\theta L(\theta) = \nabla_\theta \ln \mathrm{p}(\hat{\varphi}|f(\theta))$$

$$= \left(\nabla_\theta \otimes f^{\mathrm{T}}(\theta) \right) \Sigma_{\varphi_\epsilon}^{-1} (\hat{\varphi} - f(\theta)) = \Phi^{\mathrm{T}}(\theta) \Sigma_{\varphi_\epsilon}^{-1} (\hat{\varphi} - f(\theta)) \stackrel{!}{=} 0_M. \qquad (4.33)$$

In the remainder of this subsection, we present the specifics for the iterative Newton–Raphson and Gauss–Newton algorithms with respect to Equations (4.32) and (4.33). Although, the derivations strictly hold only for the navigation equation defined by Equation (4.6), it is very simple to modify the derivation so that we obtain the solutions for Equations (4.7), (4.9), (4.23), (4.24), or (4.26).

Newton–Raphson Algorithm

Analogously to Section 4.1.1.2, the Newton–Raphson iteration for the *weighted nonlinear least squares* problem with erroneous AOA measurements in the downlink and unknown MT orientation is given by

$$\theta^{(k+1)} = \theta^{(k)} - \left(\left(\Gamma(\theta)^{-1} \right) g(\theta) \right) \Big|_{\theta = \theta^{(k)}},$$

with the $M \times M$ Jacobian matrix

$$\Gamma(\theta) = \nabla_\theta^T \otimes g(\theta) = \nabla_\theta^T \otimes \left(\left(\nabla_\theta \otimes f^T (\theta) \right) \Sigma_{\varphi_\epsilon}^{-1} (\hat{\varphi} - f(\theta)) \right)$$

$$= \sum_{i=1}^{N} H_{f_i(\theta)} \left(\Sigma_{\varphi_\epsilon}^{-1} (\hat{\varphi} - f(\theta)) \right)_i - \Phi^T (\theta) \Sigma_{\varphi_\epsilon}^{-1} \Phi(\theta),$$

where $(\cdot)_i$ denotes the i-th element of a vector, and the $M \times M$ Hessian matrix

$$H_{f_i(\theta)} = \left(\nabla_\theta \otimes \nabla_\theta^T \right) f_i(\theta) = \begin{pmatrix} \frac{\partial^2 f_i(\theta)}{\partial \theta_1^2} & \frac{\partial^2 f_i(\theta)}{\partial \theta_1 \partial \theta_2} & \cdots & \frac{\partial^2 f_i(\theta)}{\partial \theta_1 \partial \theta_M} \\ \frac{\partial^2 f_i(\theta)}{\partial \theta_2 \partial \theta_1} & \frac{\partial^2 f_i(\theta)}{\partial \theta_2^2} & \cdots & \frac{\partial^2 f_i(\theta)}{\partial \theta_2 \partial \theta_M} \\ \vdots & & \ddots & \vdots \\ \frac{\partial^2 f_i(\theta)}{\partial \theta_M \partial \theta_1} & \cdots & \frac{\partial^2 f_i(\theta)}{\partial \theta_M \partial \theta_{M-1}} & \frac{\partial^2 f_i(\theta)}{\partial \theta_M^2} \end{pmatrix}.$$

Gauss–Newton Algorithm

Similar to Section 4.1.1.2, we now linearize Equation (4.31) at $\theta^{(k)}$:

$$f(\theta) \approx \tilde{f}(\theta) = f \left(\theta^{(k)} \right) + \Phi(\theta) \big|_{\theta = \theta^{(k)}} \left(\theta - \theta^{(k)} \right).$$

Thus, the $N \times M$ Jacobian matrix of the linearized function $\tilde{f}(\theta)$ is

$$\tilde{\Phi}(\theta) = \nabla_\theta^T \otimes \tilde{f}(\theta) = \Phi(\theta) \big|_{\theta = \theta^{(k)}}.$$

Consequently, the function

$$\tilde{g}(\theta) = \Phi^T (\theta) \big|_{\theta = \theta^{(k)}} \Sigma_{\varphi_\epsilon}^{-1} \left(\hat{\varphi} - f \left(\theta^{(k)} \right) - \Phi(\theta) \big|_{\theta = \theta^{(k)}} \left(\theta - \theta^{(k)} \right) \right)$$

is now linear in θ. Hence, setting $\tilde{g}(\theta) \overset{!}{=} 0$ and solving for θ yields the Gauss–Newton iteration

$$\theta^{(k+1)} = \theta^{(k)} + \left(\left(\Phi^T (\theta) \Sigma_{\varphi_\epsilon}^{-1} \Phi(\theta) \right)^{-1} \Phi^T (\theta) \right) \Big|_{\theta = \theta^{(k)}} \Sigma_{\varphi_\epsilon}^{-1} \left(\hat{\varphi} - f \left(\theta^{(k)} \right) \right).$$

4.2 Trilateration

In this section, we consider trilateration based on measured quantities whose values are a function of the distance between the MT and BSs. Examples of such measurements are the *TOA* measurement (see Section 3.2) and the *RSS* (see Section 2.3.2).

For TOA, the relation between the distance the signal propagates and the time is

$$d_i = c(T_i - T_0),$$

where c denotes the speed of light, T_i denotes the time instance at which the signal was received, and T_0 the time instance at which the signal was transmitted (see Section 2.1.1).

For RSS, the relationship between the received signal power and the distance depends on the propagation environment. Under ideal free space propagation conditions, the relationship is

$$d_i \propto \sqrt{\frac{P_t}{P_r}}.$$

For more details, the reader may refer to Section 2.3.2.

In the remainder of this section, we focus on trilateration with TOA measurements.

4.2.1 Trilateration with Ideal Measurements

For ideal TOA measurements, the navigation equation in two dimensions is given by (see Equation 2.1)

$$d_i = c(T_i - T_0) = \sqrt{(x - x_i)^2 + (y - y_i)^2}, \qquad i = 1, \dots, N. \qquad (4.34)$$

Squaring Equation (4.34), we obtain

$$d_i^2 = (x - x_i)^2 + (y - y_i)^2 \qquad\qquad i = 1, \dots, N, \qquad (4.35)$$

which defines N circles with center (x_i, y_i) and radius d_i. As shown in Figure 2.1, two circles defined by the TOA measurements intersect in two points, for example, BS_1 and BS_2. Only with a third circle defined by another TOA measurement, for example, BS_3, the ambiguity can be resolved. Thus, we expect that at least three independent measurements are needed in two dimensions to solve the TOA navigation equation (4.34).

The extension from two- to three-dimensional TOA-based positioning is straightforward. Instead of circles, the TOA measurements define spheres. Due to the additional unknown z-coordinate, we expect that at least four independent measurements are needed in three dimensions.

4.2.1.1 Analytical Solution to the Trilateration Equations

In this section, we derive the analytical solution to the trilateration navigation equation with ideal measurements. Given the N circles defined by Equation (4.35), we subtract the equation for $i = 1$ from the others:

$$(x - x_i)^2 + (y - y_i)^2 - (x - x_1)^2 - (y - y_1)^2 = d_i^2 - d_1^2 \qquad i = 2, \dots, N.$$

Next, we simplify this equation to obtain the linear equation system for the MT position (x, y)

$$-2x(x_i - x_1) - 2y(y_i - y_1) + x_i^2 - x_1^2 + y_i^2 - y_1^2 = d_i^2 - d_1^2 \qquad i = 2, \dots, N. \qquad (4.36)$$

We can reformulate Equation (4.36) in vector-matrix notation with $r_i^2 = x_i^2 + y_i^2$ as

$$2\underbrace{\begin{pmatrix} x_2 - x_1 & y_2 - y_1 \\ x_3 - x_1 & y_3 - y_1 \\ \vdots & \vdots \\ x_N - x_1 & y_N - y_1 \end{pmatrix}}_{=:A} \underbrace{\begin{pmatrix} x \\ y \end{pmatrix}}_{=:\theta} = \underbrace{\begin{pmatrix} r_2^2 - d_2^2 - r_1^2 + d_1^2 \\ r_3^2 - d_3^2 - r_1^2 + d_1^2 \\ \vdots \\ r_N^2 - d_N^2 - r_1^2 + d_1^2 \end{pmatrix}}_{=:B}$$

$$A\theta = B.$$

For $N = 3$, this equation yields the unique solution

$$\theta = A^{-1}B$$

given that A is invertible, that is,

$$\det A = (x_2 - x_1)(y_3 - y_1) - (x_3 - x_1)(y_2 - y_1)$$
$$= x_1(y_2 - y_3) + x_2(y_3 - y_1) + x_3(y_1 - y_2) \neq 0.$$

As expected, we obtain a unique solution with 3 TOA measurements, that is, three circles intersecting at one point (see Figure 2.1).

The extension to three dimensions is straightforward and hence, omitted for brevity.

4.2.2 Trilateration with Erroneous Measurements

4.2.2.1 Noisy Measurements

As explained in Section 2.1.1, the TOA measurements are usually corrupted by the omnipresent thermal noise. Thus, Equation (4.34) becomes

$$\hat{d}_i = d_i + \epsilon_i = c(T_i - T_0) + \epsilon_i = \sqrt{(x - x_i)^2 + (y - y_i)^2} + \epsilon_i, \quad i = 1, \dots, N. \quad (4.37)$$

Here ϵ_i denotes the *AWGN* with zero mean and standard deviation σ_{ϵ_i}. In this case, the circles do not necessarily intersect any more in one point. Instead, we can depict the noise uncertainty by a ring around the mean d_i as in Figure 2.2.

4.2.2.2 Clock Bias

In general, we cannot assume that the MT clock is synchronized to the system time T_0 of the BSs. Thus, the distance estimates in Equation (4.37) become biased *pseudo-range estimates* $\hat{d}_i$, that is,

$$\hat{d}_i = c(T_i - T_0) + \underbrace{c(T_0 - T_M)}_{=:b} + \epsilon_i = d_i + b + \epsilon_i$$

$$= \sqrt{(x - x_i)^2 + (y - y_i)^2} + b + \epsilon_i \qquad i = 1, \dots, N, \qquad (4.38)$$

where T_M, b, and $\hat{d}_i$ denote the local MT time, the clock bias between the reference time T_0 and T_M, and the pseudo-range including the effect of the clock bias. The navigation equation (4.38) is linear in the clock bias, but nonlinear in the coordinates x and y. One way of interpreting Equation (4.38) is to consider b as an uncertainty to the true radius d_i. In Figure 4.6, we plot the corresponding solution geometries. The horizontal plane displays the x- and y-coordinates whereas the vertical axis displays the clock bias b. For $b = 0$, the MT position would be the intersection of the three dashed circles. Due to the clock bias $b = c(T_0 - T_M)$, the circles defined by the pseudo-range measurements $\hat{d}_i$, $i = 1, 2, 3$, do not intersect in one point. Thus, an algorithm solving Equation (4.38) needs to adjust the clock bias until the three measurements

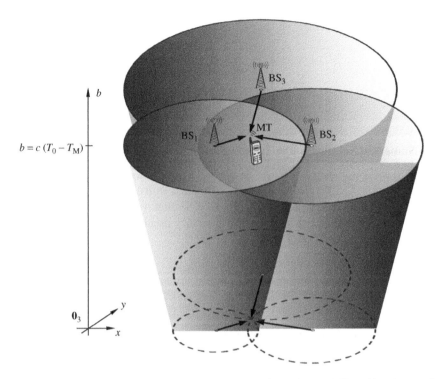

Figure 4.6 Trilateration with TOA measurements to determine unknown MT position (x, y) and clock bias b: Unknown MT clock bias yields pseudo-range measurements, which do not intersect at a single point. The solution geometry comes from the intersections of 'light' cones

intersect at one point. The solution for x, y, b can be ambiguous with only three pseudo-range measurements. Hence, an additional independent pseudo-range measurement from a fourth BS could be required to solve Equation (4.38) uniquely assuming no additional uncertainties are introduced by the noise terms ϵ_i. The solution geometry in Figure 4.6 for x, y, and b from Equation (4.38) are the intersections of 'light' cones (see Wikipedia 2013a). Each cone has a rotational symmetry axis parallel to the b-axis and through the BS position. At $b = c(T_0 - T_M)$, the radius of the cone is $\hat{d}_i$ for BS_i.

In the following, we rewrite Equation (4.38) in vector-matrix notation. For that, we collect all pseudo-ranges $\hat{d}_i$ in the vector $\hat{d} = (\hat{d}_1, \ldots , \hat{d}_N)^T$, all true ranges d_i between BSs and the MT in the vector $d = (d_1, \ldots , d_N)^T$, and the noise terms ϵ_i for all BSs in the vector $\epsilon = (\epsilon_1, \ldots , \epsilon_N)^T$. Then, we can write Equation (4.38) in vector notation as

$$\hat{d} = d + b\mathbf{1}_N + \epsilon,$$

where $\mathbf{1}_N$ is the N-dimensional vector containing only ones. For the noise vector ϵ, we define the correlation matrix

$$\Sigma_\epsilon = E\left\{\epsilon\epsilon^T\right\}. \tag{4.39}$$

The matrix Σ_ϵ takes into account all errors such as noise and residual errors, for example, after multipath mitigation (see Section 7.3). Assuming that the link-level 'TOA' measurements are independent with covariances $\sigma_{\epsilon_1}^2, \ldots, \sigma_{\epsilon_N}^2$, Equation (4.39) becomes a diagonal matrix

$$\Sigma_\epsilon = \begin{pmatrix} \sigma_{\epsilon_1}^2 & 0 & \cdots & 0 \\ 0 & \sigma_{\epsilon_2}^2 & \ddots & \vdots \\ \vdots & \ddots & \ddots & 0 \\ 0 & \cdots & 0 & \sigma_{\epsilon_N}^2 \end{pmatrix}. \tag{4.40}$$

4.2.2.3 Numerical Solution to the Trilateration Equations

For erroneous measurements, we chose an approach similar to the ML TOA estimation in AWGN (see Section 3.2.2) or the ML position estimation with erroneous AOA measurements in the downlink and unknown MT orientation (see Section 4.1.2.2). Given Equation (4.38) and (4.39), we can write the log-likelihood of the noisy pseudorange measurements $\hat{d}$ given the true distance d and the clock bias b as

$$\ln p\left(\hat{d}\middle| d\,(x),b\right) = -\frac{N}{2}\ln\,(2\pi) - \frac{1}{2}\ln\det\Sigma_\epsilon$$
$$-\frac{1}{2}\left(\left(\hat{d}-d\,(x)-b\mathbf{1}_N\right)^{\mathrm{T}}\Sigma_\epsilon^{-1}\left(\hat{d}-d\,(x)-b\mathbf{1}_N\right)\right).$$

As only the last term in the previous equation depends on the unknown MT position x and the MT clock bias b, we define the cost function

$$L\,(x,b) = -\frac{1}{2}\left(\hat{d}-d\,(x)-b\mathbf{1}_N\right)^{\mathrm{T}}\Sigma_\epsilon^{-1}\left(\hat{d}-d\,(x)-b\mathbf{1}_N\right), \tag{4.41}$$

Equation (4.41) defines the *weighted nonlinear least squares* problem for trilateration. The *ML estimate* can be found by computing the gradient of the cost function $L\,(x,b)$ and setting it equal to zero (Kay 1993):

$$\nabla_{x,b}L\,(x,b) = \nabla_{x,b}\ln p\left(\hat{d}\middle| d\,(x),b\right)$$
$$= \left(\nabla_{x,b}\otimes\left(d\,(x)+b\mathbf{1}_N\right)^{\mathrm{T}}\right)\Sigma_\epsilon^{-1}\left(\hat{d}-d\,(x)-b\mathbf{1}_N\right) = \mathbf{0}_3, \tag{4.42}$$

where $\nabla_{x,b} = \left(\frac{\partial}{\partial x}, \frac{\partial}{\partial y}, \frac{\partial}{\partial b}\right)^{\mathrm{T}}$ and $\mathbf{0}_3 = (0,0,0)^{\mathrm{T}}$. For uncorrelated TOA measurements, Equation (4.42) simplifies to

$$\nabla_{x,b}L\,(x,b) = \sum_{i=0}^{N}\begin{pmatrix} \frac{x-x_i}{d_i(x)} \\ \frac{y-y_i}{d_i(x)} \\ 1 \end{pmatrix}\frac{\hat{d}_i-d_i(x)-b}{\sigma_{\epsilon_i}^2} = \mathbf{0}_3, \tag{4.43}$$

Then, the ML estimate is the solution of Equation (4.42) or (4.43) that minimized the cost function in Equation (4.41)

$$\begin{pmatrix} \hat{x} \\ \hat{b} \end{pmatrix} = \underset{x,b}{\arg\min}\,L\,(x,b).$$

In general, there exists no analytical solution to this nonlinear three-dimensional problem. Hence, we have to apply iterative algorithms.

Gauss–Newton Algorithm

A standard algorithm to minimize the *weighted nonlinear least squares* cost function $L(x, b)$ is the Gauss–Newton algorithm (Foy 1976). The Gauss–Newton algorithm linearizes the system model about some initial value $x^{(0)}$ yielding

$$d(x) \approx d\left(x^{(0)}\right) + \Phi(x)|_{x=x^{(0)}}\left(x - x^{(0)}\right)$$

with the elements of the $N \times 3$ Jacobian matrix

$$\Phi(x) = \nabla_{x,b}^{\mathrm{T}} \otimes d(x) = \begin{pmatrix} \frac{x-x_1}{d_1} & \frac{y-y_1}{d_1} & 1 \\ \frac{x-x_2}{d_2} & \frac{y-y_2}{d_2} & 1 \\ \vdots & \vdots & \vdots \\ \frac{x-x_N}{d_N} & \frac{y-y_N}{d_N} & 1 \end{pmatrix}.$$

Next, the standard linear least squares method (Kay 1993) is applied resulting in the iterated solution

$$\begin{pmatrix} \hat{x}^{(k+1)} \\ \hat{b}^{(k+1)} \end{pmatrix} = \begin{pmatrix} \hat{x}^{(k)} \\ \hat{b}^{(k)} \end{pmatrix} + \underbrace{\left(\Phi^{\mathrm{T}}\left(\hat{x}^{(k)}\right) \Sigma_\epsilon^{-1} \Phi\left(\hat{x}^{(k)}\right)^{-1}\right)}_{A^{(k),-1}} \cdot$$

$$\underbrace{\Phi^{\mathrm{T}}\left(\hat{x}^{(k)}\right) \Sigma_\epsilon^{-1}\left(\hat{d} - d\left(\hat{x}^{(k)}\right) - b^{(k)}\mathbf{1}_N\right)}_{=:g^{(k)}}$$

$$= \begin{pmatrix} \hat{x}^{(k)} \\ \hat{b}^{(k)} \end{pmatrix} + A^{(k),-1}g^{(k)}.$$

The Gauss–Newton algorithm provides very fast convergence and accurate estimates for good initial values. For example, we may choose as initial value

$$\begin{pmatrix} \hat{x}^{(0)} \\ \hat{b}^{(0)} \end{pmatrix} = \mathbf{0}_3,$$

if the origin of the coordinate system is close to the measured BSs. Alternative, we may chose the mean position of the measured BSs, that is, $\hat{x}^{(0)} = 1/N \sum_{i=1}^{N} x_i$. For poor initial values and bad geometric conditions the algorithm results in a rank-deficient, and thus, non-invertible matrix $A^{(k)}$ for certain BS and MT constellations. In this case the algorithm diverges. Thus, we next present two more robust algorithms.

Steepest Descent Algorithm

Contrary to Gauss–Newton, the steepest descent algorithm (Kay 1993) is gradient based procedure with search direction $\nabla_{x,b} = L(x, b)$ and step size μ yielding

$$\begin{pmatrix} \hat{x}^{(k+1)} \\ \hat{b}^{(k+1)} \end{pmatrix} = \begin{pmatrix} \hat{x}^{(k)} \\ \hat{b}^{(k)} \end{pmatrix} + \mu^{(k)}\underbrace{\Phi^{\mathrm{T}}\left(\hat{x}^{(k)}\right) \Sigma_\epsilon^{-1}\left(\hat{d} - d\left(\hat{x}^{(k)}\right) - b^{(k)}\mathbf{1}_N\right)}_{=:g^{(k)}}$$

$$= \begin{pmatrix} \hat{x}^{(k)} \\ \hat{b}^{(k)} \end{pmatrix} + \mu^{(k)}g^{(k)}.$$

The easiest way to find a step size is to choose a constant $\mu^{(k)} = \mu$ for all iterations. Alternatively, a line-search procedure can be performed to find the optimum step-size per iteration. The main drawbacks of the steepest descent method are the possibility to run in local minima and the slow convergence in the final iterations.

Levenberg–Marquardt Algorithm

To cope with the problems of Gauss–Newton and steepest descent (robustness and slow convergence), a method introduced by Levenberg and Marquardt (Levenberg 1944; Marquardt 1963) is adapted to the positioning problem (Mensing and Plass 2006). It is based on a damped Gauss–Newton procedure given by

$$\begin{pmatrix} \hat{\mathbf{x}}^{(k+1)} \\ \hat{b}^{(k+1)} \end{pmatrix} = \begin{pmatrix} \hat{\mathbf{x}}^{(k)} \\ \hat{b}^{(k)} \end{pmatrix} + \underbrace{\left(\boldsymbol{\Phi}^T \left(\hat{\mathbf{x}}^{(k)} \right) \boldsymbol{\Sigma}_\epsilon^{-1} \boldsymbol{\Phi} \left(\hat{\mathbf{x}}^{(k)} \right) + \lambda^{(k)} \mathbf{I}_3 \right)^{-1}}_{\mathbf{A}^{(k)}} \cdot$$

$$\underbrace{\boldsymbol{\Phi}^T \left(\hat{\mathbf{x}}^{(k)} \right) \boldsymbol{\Sigma}_\epsilon^{-1} \left(\hat{\mathbf{d}} - \mathbf{d} \left(\hat{\mathbf{x}}^{(k)} \right) - b^{(k)} \mathbf{1}_N \right)}_{=: \mathbf{g}^{(k)}}$$

$$= \begin{pmatrix} \hat{\mathbf{x}}^{(k)} \\ \hat{b}^{(k)} \end{pmatrix} + \left(\mathbf{A}^{(k)} + \lambda^{(k)} \mathbf{I}_3 \right)^{-1} \mathbf{g}^{(k)}.$$

The damping parameter $\lambda^{(k)}$ makes sure that the appropriate matrix–in comparison to Gauss–Newton–can always be inverted yielding a much more robust implementation. The damping parameter can be calculated using a computational efficient algorithm that is based on a suboptimum line search procedure (Mensing and Plass 2006). Note that for $\lambda^{(k)} = 0$, the Levenberg–Marquardt algorithm yields the Gauss–Newton solution whereas for $|\lambda^{(k)}| \gg 1$ the Levenberg–Marquardt algorithm yields the steepest descent solution. The Levenberg–Marquardt algorithm provides fast convergence and is very robust against the initial value problem of the Gauss–Newton.

4.3   Multilateration

In this section, we examine multilateration based on measured quantities whose values are a function of the distance difference $d_i - d_p$ of the two distances between the MT and BS$_i$ and the MT and BS$_p$. Examples of such measurements are the *TDOA* measurement (see Section 2.1.2) or the *Frequency Difference of Arrival (FDOA)* measurement (Ho and Chan 1997). In the following, we will focus only on TDOA measurements as these are widely used in wireless communications systems (see Chapter 8).

4.3.1   Multilateration with Ideal Measurements

For ideal TDOA measurements, the navigation equation in two dimensions is given by (see Equation 2.4)

$$\Delta d_{i,p} = d_i - d_p = c(T_i - T_0) - c(T_p - T_0) = c(T_i - T_p) = c \cdot \Delta T_{i,p}$$
$$i \neq p, \quad i = 1, \dots, N. \qquad (4.44)$$

Without loss of generality, we choose $p = 1$. As discussed in Section 2.1.2, the time difference in Equation (4.44) removes the dependence on the system time T_0. Thus, it is not required that the MT clock is synchronized to the system time, that is, $T_0 \neq T_M$ or equivalent the clock bias b (see Section 4.2.2.2) can be unknown. Further, the TDOA navigation equation (4.44) contains only $N - 1$ equations compared to the N equations of the TOA navigation equation (see Equation (4.34), (4.37), or (4.38)). The solution geometry of Equation (4.44) are hyperbolas in two dimensions as demonstrated in Section 2.1.2.

4.3.1.1 Analytic solution to the Multilateration Equations

In the following, we derive the analytical solution to the multilateration navigation equation with ideal measurements. First, we take the first equality in Equation (4.44), solve it for d_i and square the equation to obtain

$$d_i^2 = \left(\Delta d_{i,1} + d_1\right)^2 = \Delta d_{i,1}^2 + 2\Delta d_{i,1} d_1 + d_1^2 \qquad i = 2, \dots, N. \qquad (4.45)$$

Next, we rearrange Equation (4.45) and divide it by $\Delta d_{i,1}$:

$$0 = \Delta d_{i,1}^2 + 2\Delta d_{i,1} d_1 + d_1^2 - d_i^2$$
$$0 = \Delta d_{i,1} + 2d_1 + \frac{d_1^2 - d_i^2}{\Delta d_{i,1}} \qquad i = 2, \dots, N. \qquad (4.46)$$

Note, we assumed that $\Delta d_{i,1} \neq 0$. Next, we subtract Equation (4.46) for $i = 2$ from Equation (4.46) for $i = 3, \dots, N$

$$0 = \Delta d_{i,1} - \Delta d_{2,1} + \frac{d_1^2 - d_i^2}{\Delta d_{i,1}} - \frac{d_1^2 - d_2^2}{\Delta d_{2,1}} \qquad i = 3, \dots, N. \qquad (4.47)$$

Further, we have, according to Equation (4.36)

$$d_1^2 - d_i^2 = 2\left(x_i - x_1\right) x + 2\left(y_i - y_1\right) y + x_1^2 + y_1^2 - x_i^2 - y_i^2 \qquad i = 2, \dots, N.$$

This equation implicitly assumes that there is no clock bias, that is, $b = 0$. Plugging the previous equation into Equation (4.47), we obtain the linear equations in x and y

$$0 = \Delta d_{i,1} - \Delta d_{2,1} + \frac{x_1^2 + y_1^2 - x_i^2 - y_i^2}{\Delta d_{i,1}} - \frac{x_1^2 + y_1^2 - x_2^2 - y_2^2}{\Delta d_{2,1}}$$
$$+ 2\left(\frac{x_i - x_1}{\Delta d_{i,1}} - \frac{x_2 - x_1}{\Delta d_{2,1}}\right) x + 2\left(\frac{y_i - y_1}{\Delta d_{i,1}} - \frac{y_2 - y_1}{\Delta d_{2,1}}\right) y \qquad i = 3, \dots, N.$$

Using vector-matrix notation with $r_i^2 = x_i^2 + y_i^2$, the former equation becomes

$$
2\underbrace{\begin{pmatrix}
\left(\frac{x_3-x_1}{\Delta d_{3,1}} - \frac{x_2-x_1}{\Delta d_{2,1}}\right) & \left(\frac{y_3-y_1}{\Delta d_{3,1}} - \frac{y_2-y_1}{\Delta d_{2,1}}\right) \\
\left(\frac{x_4-x_1}{\Delta d_{4,1}} - \frac{x_2-x_1}{\Delta d_{2,1}}\right) & \left(\frac{y_4-y_1}{\Delta d_{4,1}} - \frac{y_2-y_1}{\Delta d_{2,1}}\right) \\
\vdots & \\
\left(\frac{x_N-x_1}{\Delta d_{N,1}} - \frac{x_2-x_1}{\Delta d_{2,1}}\right) & \left(\frac{y_N-y_1}{\Delta d_{N,1}} - \frac{y_2-y_1}{\Delta d_{2,1}}\right)
\end{pmatrix}}_{=:A}
\underbrace{\begin{pmatrix} x \\ y \end{pmatrix}}_{=:\theta} =
$$

$$
= \underbrace{\begin{pmatrix}
\Delta d_{3,1} - \Delta d_{2,1} + \frac{r_1^2-r_3^2}{\Delta d_{3,1}} - \frac{r_1^2-r_2^2}{\Delta d_{2,1}} \\
\Delta d_{4,1} - \Delta d_{2,1} + \frac{r_1^2-r_4^2}{\Delta d_{4,1}} - \frac{r_1^2-r_2^2}{\Delta d_{2,1}} \\
\vdots \\
\Delta d_{N,1} - \Delta d_{2,1} + \frac{r_1^2-r_N^2}{\Delta d_{N,1}} - \frac{r_1^2-r_2^2}{\Delta d_{2,1}}
\end{pmatrix}}_{=:B}
$$

$$A\theta = B.$$

For $N = 4$, we obtain the unique solution

$$\theta = A^{-1}B$$

given that A is invertible, that is, $\det A \neq 0$.

Alternative, we can use the method presented in (Figueiras and Frattasi 2010) to solve Equation (4.44). First, we rearrange Equation (4.45) to

$$(x - x_i)^2 + (y - y_i)^2 - (x - x_1)^2 - (y - y_1)^2 = \Delta d_{i,1}^2 + 2\Delta d_{i,1}d_1 \qquad i = 2, \dots, N.$$

Then, we simplify this equation using $r_i^2 = x_i^2 + y_i^2$

$$-2x(x_i - x_1) - 2y(y_i - y_1) + r_2^2 - r_1^2 = \Delta d_{2,1}^2 + 2\Delta d_{2,1}d_1 \qquad i = 2, \dots, N.$$

Next, we use vector-matrix notation to obtain the linear equation system for the unknowns x, y, and d_1

$$
2\underbrace{\begin{pmatrix}
x_2 - x_1 & y_2 - y_1 & \Delta d_{2,1} \\
x_3 - x_1 & y_3 - y_1 & \Delta d_{3,1} \\
\vdots & & \\
x_N - x_1 & y_N - y_1 & \Delta d_{N,1}
\end{pmatrix}}_{=:A}
\underbrace{\begin{pmatrix} x \\ y \\ d_1 \end{pmatrix}}_{=:\hat{\theta}} =
\underbrace{\begin{pmatrix}
r_2^2 - \Delta d_{2,1}^2 - r_1^2 \\
r_3^2 - \Delta d_{3,1}^2 - r_1^2 \\
\vdots \\
r_N^2 - \Delta d_{N,1}^2 - r_1^2
\end{pmatrix}}_{=:B}
$$

$$A\hat{\theta} = B.$$

For $N = 4$, we obtain the unique solution

$$\hat{\theta} = A^{-1}B$$

given that A is invertible, that is, $\det A \neq 0$.

4.3.2   Multilateration with Erroneous Measurements

Similar to Section 4.2.2, TDOA measurements are in general corrupted by AWGN. Assuming that the TDOA measurement is computed from two noisy TOA measurements, Equation (4.44) becomes

$$\Delta \hat{d}_{i,1} = c\Delta \hat{T}_{i,1} = c\left(\hat{T}_i - \hat{T}_1\right) = d_i - d_1 + \underbrace{\epsilon_i - \epsilon_1}_{=:\epsilon_{i,1}} = \Delta d_{i,1} + \Delta \epsilon_{i,1},$$

$$i = 2, \dots, N.$$

Note that the TDOA $\Delta \hat{d}_{i,1}$ is Gaussian distributed with mean $\Delta d_{i,1}$ and variance $\sigma^2_{\Delta \epsilon_{i,1}} = \sigma^2_{\epsilon_i} + \sigma^2_{\epsilon_1}$. We can rewrite previous equation in vector notation as

$$\Delta \hat{d}_1 = \Delta d_1 + \Delta \epsilon_1,$$

where $\Delta \hat{d}_1 = (\Delta \hat{d}_{2,1}, \dots, \Delta \hat{d}_{N,1})^{\mathrm{T}}$, $\Delta d_1 = (\Delta d_{2,1}, \dots, \Delta d_{N,1})^{\mathrm{T}}$, and $\Delta \epsilon_1 = (\Delta \epsilon_{2,1}, \dots, \Delta \epsilon_{N,1})^{\mathrm{T}}$. For notational convenience, we omit in the sequel the index of the reference BS_1 for the vectors $\Delta \hat{d}$, Δd, and $\Delta \epsilon$.

In the following, we use the notation $\epsilon_{(2:N)} = (\epsilon_2, \dots, \epsilon_N)^{\mathrm{T}}$ to denote a subvector of ϵ containing rows $2, \dots, N$. For the noise vector $\Delta \epsilon$, we define the noise correlation matrix

$$\Sigma_{\Delta \epsilon} = \mathrm{E}\left\{\Delta \epsilon \Delta \epsilon^{\mathrm{T}}\right\} = \mathrm{E}\left\{\left(\epsilon_{(2:N)} - \epsilon_1 \mathbf{1}_{N-1}\right)\left(\epsilon_{(2:N)} - \epsilon_1 \mathbf{1}_{N-1}\right)^{\mathrm{T}}\right\}$$

$$= \Sigma_{\epsilon_{(2:N)}} - \mathrm{E}\left\{\epsilon_{(2:N)}\epsilon_1 \mathbf{1}_{N-1}^{\mathrm{T}}\right\} - \mathrm{E}\left\{\epsilon_1 \mathbf{1}_{N-1}\epsilon_{(2:N)}^{\mathrm{T}}\right\} + \mathrm{E}\left\{\epsilon_1^2\right\}\mathbf{1}_{(N-1)\times(N-1)}, \quad (4.48)$$

where the noise correlation matrix $\Sigma_{\epsilon_{(2:N)}} \in \mathbb{R}^{(N-1)\times(N-1)}$ is defined by Equation (4.39). Assuming that the cross-correlation terms in Equation (4.48) are zero, that is, $\mathrm{E}\left\{\epsilon_{(2:N)}\epsilon_1\mathbf{1}_{N-1}^{\mathrm{T}}\right\} = \mathrm{E}\left\{\epsilon_1\mathbf{1}_{N-1}\epsilon_{(2:N)}^{\mathrm{T}}\right\} = \mathbf{0}_{(N-1)\times(N-1)}$ and that the link-level 'TOA' measurements are independent with covariances $\sigma^2_{\epsilon_1}, \dots, \sigma^2_{\epsilon_N}$, Equation (4.48) becomes

$$\Sigma_{\Delta \epsilon} = \Sigma_{\epsilon_{(2:N)}} + \mathrm{E}\left\{\epsilon_1^2\right\}\mathbf{1}_{(N-1)\times(N-1)}$$

$$= \begin{pmatrix} \sigma^2_{\epsilon_1} + \sigma^2_{\epsilon_2} & \sigma^2_{\epsilon_1} & \cdots & \sigma^2_{\epsilon_1} \\ \sigma^2_{\epsilon_1} & \sigma^2_{\epsilon_1} + \sigma^2_{\epsilon_3} & \cdots & \sigma^2_{\epsilon_1} \\ \vdots & \vdots & \ddots & \vdots \\ \sigma^2_{\epsilon_1} & \sigma^2_{\epsilon_1} & \cdots & \sigma^2_{\epsilon_1} + \sigma^2_{\epsilon_N} \end{pmatrix}. \quad (4.49)$$

As in Section 4.2.2, the matrix $\Sigma_{\Delta \epsilon}$ takes into account all errors such as noise and residual errors, for example, after multipath mitigation (see Section 7.3).

4.3.2.1   Numerical Solution to the Multilateration Equations

Equivalent to Section 4.2.2.3, we can solve the navigation equation with the *weighted nonlinear least squares* approach minimizing the cost function

$$L(x) = \left(\Delta \hat{d} - \Delta d(x)\right)^{\mathrm{T}} \Sigma_{\Delta \epsilon}^{-1} \left(\Delta \hat{d} - \Delta d(x)\right).$$

with respect to the unknown MT position. Here, $\Delta\hat{d} \in \mathbb{R}^{N-1}$ includes the $N-1$ measured TDOAs and $\Delta d(x)$ the ideal, real TDOAs. As before, a suitable algorithm provides the MT position estimate

$$\hat{x} = \arg\min_{x} L(x).$$

In general, there exists no analytical solution to this nonlinear two or three-dimensional problem. Hence, we have to apply iterative algorithms.

Gauss–Newton Algorithm

As for solving the trilateration navigation equations, we can apply the Gauss–Newton algorithm (Foy 1976) to solve the multilateration navigation equations. In this case, the linearization step is given by

$$\Delta d(x) \approx \Delta d\left(x^{(0)}\right) + \Phi(x)|_{x=x^{(0)}}\left(x - x^{(0)}\right)$$

with the elements of the $(N-1)\times 2$ Jacobian matrix

$$\Phi(x) = \nabla_x^T \otimes \Delta d(x) = \begin{pmatrix} \frac{x-x_2}{d_2} - \frac{x-x_1}{d_1} & \frac{y-y_2}{d_2} - \frac{y-y_1}{d_1} \\ \frac{x-x_3}{d_3} - \frac{x-x_1}{d_1} & \frac{y-y_3}{d_3} - \frac{y-y_1}{d_1} \\ \vdots & \vdots \\ \frac{x-x_N}{d_N} - \frac{x-x_1}{d_1} & \frac{y-y_N}{d_N} - \frac{y-y_1}{d_1} \end{pmatrix},$$

where $\nabla_x = \left(\frac{\partial}{\partial x}, \frac{\partial}{\partial y}\right)^T$. Similar to Section 4.2.2.3, the linear least squares method is applied resulting in the iterated Gauss–Newton algorithm

$$\hat{x}^{(k+1)} = \hat{x}^{(k)} + \underbrace{\left(\Phi^T\left(\hat{x}^{(k)}\right)\Sigma_{\Delta\epsilon}^{-1}\Phi\left(\hat{x}^{(k)}\right)^{-1}\right)}_{A_{\mathrm{TDOA}}^{(k),-1}}\underbrace{\Phi^T\left(\hat{x}^{(k)}\right)\Sigma_{\Delta\epsilon}^{-1}\left(\Delta\hat{d} - \Delta d\left(\hat{x}^{(k)}\right)\right)}_{=:g_{\mathrm{TDOA}}^{(k)}}$$

$$= \hat{x}^{(k)} + A_{\mathrm{TDOA}}^{(k),-1}g_{\mathrm{TDOA}}^{(k)}.$$

Note that the noise correlation matrix for TDOAs $\Sigma_{\Delta\epsilon}$ in this equation always contains non-zero off-diagonal entries (see Equations 4.48 and 4.49) in contrast to the noise correlation matrix for TOAs (see Equations 4.39 and 4.40). Thus, the matrix $A_{\mathrm{TDOA}}^{(k)}$ can become singular for bad geometric conditions or bad initial values. Then, the Gauss–Newton algorithm can have convergence problems due to such an ill-conditioned matrix $A_{\mathrm{TDOA}}^{(k)}$. For these scenarios, the hyperbolic character of the TDOA measurements needs more robust approaches. Two of them are presented next.

Steepest Descent Algorithm

In contrast to Gauss–Newton, the gradient based steepest descent algorithm with search direction $\nabla_x = L(x)$ and step size μ yields

$$\hat{x}^{(k+1)} = \hat{x}^{(k)} + \mu^{(k)}\underbrace{\Phi^T\left(\hat{x}^{(k)}\right)\Sigma_{\Delta\epsilon}^{-1}\left(\Delta\hat{d} - \Delta d\left(\hat{x}^{(k)}\right)\right)}_{=:g_{\mathrm{TDOA}}^{(k)}}$$

$$= \hat{x}^{(k)} + \mu^{(k)}g_{\mathrm{TDOA}}^{(k)}.$$

The step size can be either chosen constant $\mu^{(k)} = \mu$ for all iterations or a line-search procedure can be performed to find the optimum step size per iteration. As in Section 4.2.2.3, the main drawbacks of the steepest descent method are the possibility of running in local minima and slow convergence in the final iterations.

Levenberg–Marquardt Algorithm

To cope with the problems of Gauss–Newton and steepest descent (robustness and slow convergence), the Levenberg–Marquardt algorithm is adapted to the TDOA-based positioning problem (Mensing and Plass 2006), that is,

$$\hat{x}^{(k+1)} = \hat{x}^{(k)} + \underbrace{\left(\mathbf{\Phi}^{\mathrm{T}} \left(\hat{x}^{(k)} \right) \mathbf{\Sigma}_{\Delta\epsilon}^{-1} \mathbf{\Phi} \left(\hat{x}^{(k)} \right) + \lambda^{(k)} I_2 \right)^{-1}}_{A_{\mathrm{TDOA}}^{(k)}} \cdot$$

$$\underbrace{\mathbf{\Phi}^{\mathrm{T}} \left(\hat{x}^{(k)} \right) \mathbf{\Sigma}_{\Delta\epsilon}^{-1} \left(\Delta\hat{d} - \Delta d \left(\hat{x}^{(k)} \right) \right)}_{=:g_{\mathrm{TDOA}}^{(k)}}$$

$$= \hat{x}^{(k)} + \left(A_{\mathrm{TDOA}}^{(k)} + \lambda^{(k)} I_2 \right) - 1 g_{\mathrm{TDOA}}^{(k)}.$$

Similar to TOA in Section 4.2.2.3, the damping parameter $\lambda^{(k)}$ ensures that the appropriate matrix – in comparison to Gauss–Newton – can always be inverted yielding a much more robust implementation.

4.4   Fingerprinting

In Section 2.3, we introduced the fingerprinting positioning method. We defined a fingerprint in wireless positioning as a set of measurable signal characteristics that depend on the position of transmission or reception. Thus, the previously considered AOA, AOAD, RSS, TOA, and TDOA measurements are wireless fingerprints. In general, fingerprinting also considers measurements that cannot be expressed through explicit functional relations between the location dependent measurements and the MT position. Therefore, fingerprinting usually builds up a database, which contains the fingerprints, that is, the set of signal characteristics for each position in the environment. The database is often created in an off-line phase. During the online phase, a fingerprinting method then compares a measured fingerprint to the fingerprint entries of that database in order to determine the database fingerprint that matches best and thus, the corresponding MT position. Consequently, the fingerprinting database can be seen as a generalization of the previously presented navigation equations.

Many different algorithms for fingerprinting exist (Liu et al. 2007a), which can be classified as probabilistic, k-nearest neighbor, neural network, support vector machine, or smallest M-vertex polygon algorithms.

The probabilistic algorithms determine for each measurement in the database and for the actual measurement a probability that the measured fingerprint corresponds to a specific location. Then, the algorithm chooses the location with the highest probability.

The k-nearest neighbor algorithms use the online measured fingerprint to determine from the database k closest matching fingerprints. From these fingerprints, the algorithm then estimates an average position.

Neural network algorithms learn during the off-line phase the implicit relation between the fingerprints and the measured locations by obtaining appropriate weights. During the online phase these weights are applied to the measured fingerprints to estimate the MT position.

For further details on fingerprinting algorithms, we refer the interested reader to (Liu et al. 2007a) and references therein.

4.5 Performance Bounds and Measures

This section presents some performance bounds and measures to asses the accuracy of the previously presented positioning algorithms. Specifically, we consider the positioning *root mean square error (RMSE)*, the *cumulative distribution function (CDF)* and the *circular error probability (CEP)* of the positioning error, the *positioning CRLB*, the *dilution of precision (DOP)*, and the *complexity*.

4.5.1   Root Mean Square Error–RMSE

The RMSE of an position estimator $\hat{x}$ is defined as

$$\epsilon_{\text{RMSE}} = \sqrt{\text{E}\left\{||x - \hat{x}||_2^2\right\}}.$$

Note that the RMSE takes into account both the variance and the bias of an estimator, that is,

$$\epsilon_{\text{RMSE}}^2 = \text{VAR}\{\hat{x}\} + ||\mathbf{Bias}\,(\hat{x})||_2^2 \qquad \text{with}$$

$$\text{VAR}\{\hat{x}\} = \text{E}\left\{||\hat{x} - \text{E}\{\hat{x}\}||_2^2\right\} \qquad \text{and} \qquad \mathbf{Bias}\,(\hat{x}) = \text{E}\{\hat{x}\} - x.$$

Thus, we can compare through the RMSE both biased and unbiased estimators.

4.5.2   Cumulative Distribution Function – CDF

With the CDF, usually outage probabilities can be analyzed, that is, at what probability will a positioning method fail to achieve a certain accuracy. The CDF of the positioning error in [m] is defined as the probability that the absolute two-dimensional position error is below the value 'error', that is,

$$\text{CDF}\,(\text{error}) = \text{P}\left(||x - \hat{x}||_2 \le \text{error}\right).$$

The CDF is usually obtained over several MT positions and noise realizations, for example, Section 5.3.

4.5.3   Circular Error Probability–CEP

The CEP is defined as the probability for which a measurement will fall into a circle with radius r (Gustafsson and Gunnarsson 2005)

$$\text{CEP}_{\text{p}}\,(r) = \text{P}\left(||x - \hat{x}||_2 < r\right) = \text{p}.$$

Typical CEPs are $p = 67\%$ and $p = 95\%$.

4.5.4   Positioning Cramér–Rao Lower Bound–CRLB

The CRLB for unbiased position estimators defines a performance lower bound on the variance of position estimation algorithms (Kay 1993). We define the *covariance matrix* of the estimator $\hat{\theta}$ as

$$\text{COV}\{\hat{\theta}\} := \text{E}_{y|\theta}\left\{\left(\hat{\theta} - \text{E}_{y|\theta}\{\hat{\theta}\}\right)\left(\hat{\theta} - \text{E}_{y|\theta}\{\hat{\theta}\}\right)^{\text{T}}\right\}.$$

The parameter vector $\theta \in \mathbb{R}^M$ includes the unknown parameters, for example, the position coordinates x and y for two-dimensional positioning and additionally z for three-dimensional positioning as well as other unknown parameters such as the MT orientations φ and ϑ (see Sections 4.1.1.2 and 4.1.1.3) or the clock bias b (see Section 4.2.2.2). For the erroneous measurement vector $y \in \mathbb{R}^N$ given the parameter vector θ, the *Fisher information matrix (FIM)* is defined as

$$J_{\text{FIM}} = \text{E}_{y|\theta}\left\{\left(\nabla_\theta \ln p(y|\theta)\right)\left(\nabla_\theta^{\text{T}} \ln p(y|\theta)\right)\right\} = -\text{E}_{y|\theta}\left\{\left(\nabla_\theta \otimes \nabla_\theta^{\text{T}}\right) \ln p(y|\theta)\right\}. \tag{4.50}$$

Then, the covariance matrix of the position estimator $\hat{x} = g(\theta)$ can be lower bounded by

$$\text{COV}\{\hat{x}\} \geq \left(\nabla_\theta^{\text{T}} \otimes g(\theta)\right)J^{-1}(\theta)\left(\nabla_\theta^{\text{T}} \otimes g(\theta)\right). \tag{4.51}$$

Note that $\geq$ in this equation does not mean that each element of the matrix on the left side is larger than or equal to each element of the matrix on the right side, but that the difference between both matrices is *positive semidefinite* (Kay 1993)

$$a^{\text{T}}\left(\text{COV}\{\hat{x}\} - \left(\nabla_\theta^{\text{T}} \otimes g(\theta)\right)J^{-1}(\theta)\left(\nabla_\theta^{\text{T}} \otimes g(\theta)\right)\right)a \geq 0 \qquad \forall a \in \mathbb{R}^M.$$

If $x = \theta$ and $g(\theta)$ is an *unbiased estimator*, i.e. $\text{E}\{g(\theta)\} = \theta$, the Jacobian matrix in Equation (4.51) becomes

$$\left(\nabla_\theta^{\text{T}} \otimes g(\theta)\right) = I_{M \times M}.$$

For a *biased estimator* $g(\theta)$, the Jacobian matrix becomes

$$\left(\nabla_\theta^{\text{T}} \otimes g(\theta)\right) = I_{M \times M} + \nabla_\theta^{\text{T}} \otimes \text{Bias}(\theta).$$

In many cases, we assume that the measurement vector y is Gaussian distributed with mean $\mu(\theta)$ and covariance matrix $\Sigma(\theta)$, that is, $y \sim \mathcal{N}(\mu(\theta), \Sigma(\theta))$. Then, Equation (4.50) simplifies to (Kay 1993)

$$J_{\text{FIM}} = \left(\nabla_\theta^{\text{T}} \otimes \mu(\theta)\right)^{\text{T}} \Sigma(\theta)^{-1}\left(\nabla_\theta^{\text{T}} \otimes \mu(\theta)\right) + \frac{1}{2}\text{trace}(D(\theta))$$

with $(D(\theta))_{ij} = \Sigma(\theta)^{-1}\frac{\partial\Sigma(\theta)}{\partial\theta_i}\Sigma(\theta)^{-1}\frac{\partial\Sigma(\theta)}{\partial\theta_j}$.

If the covariance matrix is not a function of θ, $D(\theta)$ is zero. With $x = \theta$ and $\Phi(x) = \nabla_\theta^{\text{T}} \otimes \mu(\theta)$, we can lower bound the variance of a position estimator as

$$\text{VAR}\{\hat{x}\} \geq \text{CRLB}(x) = \sqrt{\text{trace}\left(\Phi^{\text{T}}(x)\Sigma^{-1}\Phi(x)\right)^{-1}}. \tag{4.52}$$

Here, x, $\Phi(x)$, and Σ denote the M-dimensional position vector, the $N \times M$ Jacobian matrix of the measurement model, which takes into account the geometry of the BSs and MT, and the $N \times N$ noise covariance matrix of the measurements. Thus, the CRLB takes into account the geometry of the considered scenario, the measurement accuracies, and the measurement dependencies.

Figure 4.7 shows the positioning CRLB for trilateration (TOA) without and with clock bias, multilateration (TDOA), and triangulation (AOA). We calculated the CRLB according to (4.52) for TOA and AOA measurements with $\sigma_{\epsilon_i}^2 = 1$ m and $\sigma_{\varphi_{\epsilon_i}}^2 = 0.01°$. For triangulation, we analyze the downlink AOA with unknown MT orientation according to Equations (4.6), (4.7), and (4.9). The maximum plotted variance of the estimated position is 3 m. Comparing the different methods, clearly the trilateration with TOAs and no clock bias provides the best accuracy for this scenario. Both TOA with clock bias and TDOA provide only a good accuracy within the triangle formed by the three BSs. Note that the CRLBs for TOA and TDOA are the same in accordance with (Urruela et al. 2006) although the gray shading is not exactly the same due to some numerical effects along the x-axis for TDOA. To achieve a similar performance with the triangulation methods as with TDOA, we have to choose $\sigma_{\varphi_{\epsilon_i}}^2 = 0.01°$. The CRLBs for downlink AOA and AOAD with unknown MT orientation and the nonlinear arctan function from Equations (4.6) and (4.7) are the same similar to TOA with clock bias and TDOA. In contrast the CRLB for downlink AOAD with unknown MT orientation and the arccos from Equation (4.9) is different due to the different nonlinear functions.

4.5.5 Dilution of Precision–DOP

The DOP is strongly related to the CRLB. With DOP the geometric influence of the positioning scenario on the overall positioning performance is described. For trilateration with unknown MT clock bias, the matrix

$$\Xi = \left(\Phi^T(x)\Phi(x)\right)^{-1} = \begin{pmatrix} EDOP^2 & \cdot & \cdot & \cdot \\ \cdot & NDOP^2 & \cdot & \cdot \\ \cdot & \cdot & VDOP^2 & \cdot \\ \cdot & \cdot & \cdot & TDOP^2 \end{pmatrix}$$

can be separated in different DOP parts: *east DOP (EDOP)*, *north DOP (NDOP)*, *vertical DOP (VDOP)*, and *time DOP (TDOP)*. Hence, the *position DOP (PDOP)* is defined as

$$PDOP = \sqrt{EDOP^2 + NDOP^2 + VDOP^2},$$

and the overall DOP or *geometric DOP (GDOP)* is defined as

$$DOP = GDOP = \sqrt{PDOP^2 + TDOP^2} = \sqrt{\text{trace}(\Xi)}.$$

The DOP analysis can help to estimate the expected RMSE if the measurement accuracy on all links is assumed equal. Then, the DOP is due to geometrical properties of the current scenario. For instance, Figure 4.8 compares the PDOP of different positioning methods and measurement types for the geometry of Example 4.1.1. Similar to the CRLBs in Figure 4.7, the trilateration method for TOA without clock bias shows the best performance. For the PDOP, the performance of TDOA is better than that of TOA with clock bias in Figure 4.8, which is

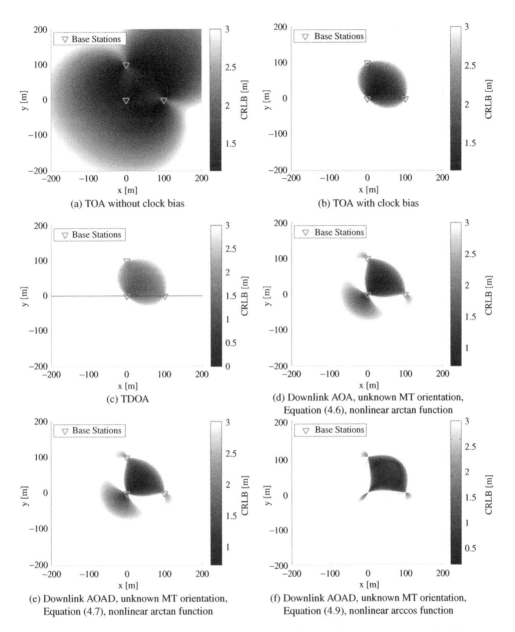

(a) TOA without clock bias

(b) TOA with clock bias

(c) TDOA

(d) Downlink AOA, unknown MT orientation, Equation (4.6), nonlinear arctan function

(e) Downlink AOAD, unknown MT orientation, Equation (4.7), nonlinear arctan function

(f) Downlink AOAD, unknown MT orientation, Equation (4.9), nonlinear arccos function

Figure 4.7 Positioning CRLB of different positioning methods and measurement types for the geometry of Example 4.1.1, $\sigma_{\epsilon_i}^2 = 1$ m for TOA measurements and $\sigma_{\varphi_{\epsilon_i}}^2 = 0.01°$ for AOA measurements

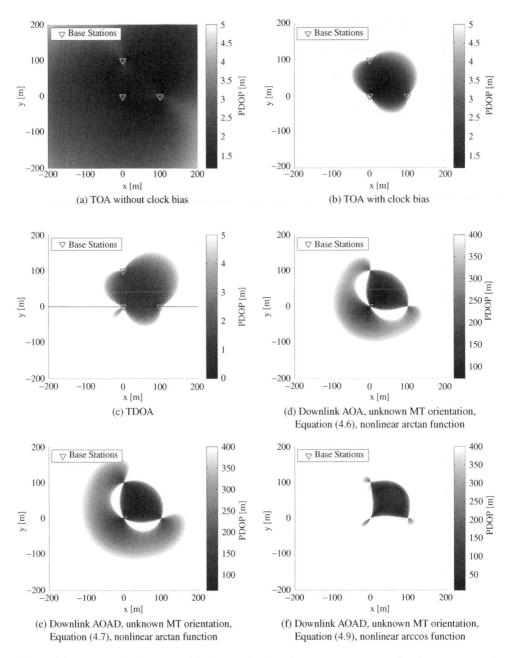

(a) TOA without clock bias

(b) TOA with clock bias

(c) TDOA

(d) Downlink AOA, unknown MT orientation,
Equation (4.6), nonlinear arctan function

(e) Downlink AOAD, unknown MT orientation,
Equation (4.7), nonlinear arctan function

(f) Downlink AOAD, unknown MT orientation,
Equation (4.9), nonlinear arccos function

Figure 4.8 PDOP of different positioning methods and measurement types for the geometry of
Example 4.1.1

in contrast to the corresponding CRLBs. As each TDOA for Figure 4.7 has a higher variance due to the subtraction of two TOAs (see Equations 4.40 and 4.49), the PDOP for TDOA must have a higher accuracy than the one for TOA so that the corresponding CRLBs for multilateration with TDOAs and trilateration with TOAs and clock bias are both the same (Urruela et al. 2006). The PDOP for triangulation (see Figures 4.8f) is two orders of magnitude larger than the one for trilateration or multilateration (see Figures 4.8c). Similar to TDOA and TOA with clock bias, the triangulation method for downlink AOAD with unknown MT orientation and the nonlinear arctan function (see Equation 4.7) is better than the one for downlink AOA (see Equation 4.6).

4.5.6   Complexity

Another way to analyze the performance of different position estimation methods is to compare their complexity. Most often this means analyzing the computational complexity of the method. However, it could also include hardware complexity – an antenna array for AOA-based positioning methods versus a single omni-directional antenna for RSS-, TOA-, or TDOA-based methods – or the additional complexity and effort due to the off-line calibration phase in fingerprinting methods.

5

Position Tracking

The approaches presented in the previous chapter consider a static solution of the position estimation. It was assumed that the MT did not move during the estimation process, and therefore, the MT position was treated as a deterministic parameter. In reality, the *MT positions* are usually *correlated over time*. For instance, considering a pedestrian user or a moving car, certain information about the position can be derived using the *history of past estimates* and suitable *movement or mobility models*. This includes restricted movements of the MT. For instance, a pedestrian cannot 'jump' from one position to another in limited time or a car usually can change its direction only 'smoothly'. This behavior can be used as side information for position tracking algorithms.

For the derivation of the algorithms (also see Arulampalam et al. 2002; Krach 2010; Mensing 2013; Ristic et al. 2004), we assume that the time axis is divided in to discrete time intervals. Further, we presume a causal system, that is, future states (such as position or velocity) cannot impact current and past estimates. However, since the past states can impact the current and future states, this property has to be reflected in the chosen model. A commonly used model in this position tracking context is a *first order hidden Markov model*.

Figure 5.1 depicts such a Markov model, for example, (Krach et al. 2008), with *unknown states* $s_k \in \mathbb{R}^{N_s}$ that have to be estimated in each time-step $k \in \mathbb{N}$. It is a hidden Markov model since the states can only be observed implicitly in terms of the available measurements or observations. The estimation process takes into account these measurements $y_k \in \mathbb{R}^{N_y}$ in each time-step k in addition to the model parameters. According to Figure 5.1, the measurements y_k depend only on the state vector s_k at the current time-step. This dependence is defined by the so-called *measurement model*

$$y_k = g_k\left(s_k, n_k\right). \tag{5.1}$$

The function

$$g_k : \mathbb{R}^{N_s} \times \mathbb{R}^{N_n} \to \mathbb{R}^{N_y}$$

is a possibly nonlinear function of the state s_k and the measurement noise $n_k \in \mathbb{R}^{N_n}$ (see Arulampalam et al. 2002). The properties of the measurement noise n_k define the measurement uncertainties. Another equivalent representation of the measurement model is based on the conditional PDF of the measurements given the states, that is, $p\left(y_k | s_k\right)$.

Positioning in Wireless Communications Systems, First Edition. Stephan Sand, Armin Dammann and Christian Mensing.
© 2014 John Wiley & Sons, Ltd. Published 2014 by John Wiley & Sons, Ltd.

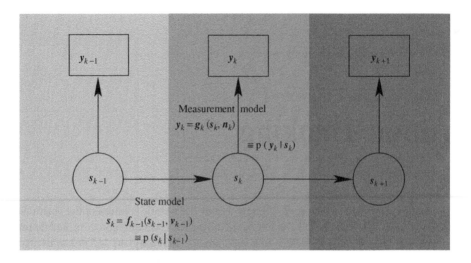

Figure 5.1   First order hidden Markov model

The *state model* defines a relationship between the previous state s_{k-1} and the current state s_k. It is given as

$$s_k = f_{k-1}\left(s_{k-1}, v_{k-1}\right),\tag{5.2}$$

where the function

$$f_{k-1} : \mathbb{R}^{N_s} \times \mathbb{R}^{N_v} \to \mathbb{R}^{N_s}$$

is a possibly nonlinear function of the state s_{k-1} and the state process noise $v_{k-1} \in \mathbb{R}^{N_v}$. The properties of the state process noise v_{k-1} define how random the state changes can be. The equivalent representation of the state model is based on the conditioned PDF $p\left(s_k|s_{k-1}\right)$. In the positioning context, the state vector can include information about the MT position or its velocity. The corresponding state model includes information about the mobility or movement of the MT. Therefore, it is often denoted as a *mobility model*.

Following the Bayesian approach, for example, (Arulampalam et al. 2002; Kay 1993; Ristic et al. 2004), it is required that the *PDF of the current state* is estimated by considering all previous and current measurements, that is, the PDF $p\left(s_k|y_1, y_2, \ \dots \ , y_k\right)$ has to be constructed. This is done recursively by assuming that the prior distribution of the state s_0 is known:

In the first step of *Bayesian estimation*, the state model is used to obtain the *prior PDF* of the state at time-step k by

$$p\left(s_k|y_1, y_2, \ \dots \ , y_{k-1}\right) = \int p\left(s_k|s_{k-1}\right) p\left(s_{k-1}|y_1, y_2, \ \dots \ , y_{k-1}\right) ds_{k-1}.\tag{5.3}$$

The PDF $p\left(s_k|s_{k-1}\right)$ is defined by the state equation and the known statistics of the state noise v_{k-1}. This step is denoted as *prediction step* since the new state is estimated as a prediction of the old state. At this stage the current measurements are not yet used.

For the second step, it is required that at time-step k the measurements y_k become available. They can be used to update the prior PDF by the Bayes' rule resulting in a normalized product of the likelihood $p\left(y_k|s_k\right)$ and the prior PDF, that is,

$$p\left(s_k|y_1,y_2, \cdots ,y_k\right) = \frac{p\left(y_k|s_k\right)p\left(s_k|y_1,y_2, \cdots ,y_{k-1}\right)}{p\left(y_k|y_1,y_2, \cdots ,y_{k-1}\right)}, \tag{5.4}$$

with normalization constant

$$p\left(y_k|y_1,y_2, \cdots ,y_{k-1}\right) = \int p\left(y_k|s_k\right)p\left(s_k|y_1,y_2, \cdots ,y_{k-1}\right)dy_k.$$

Therefore, the *posterior PDF* can be calculated by using the measurement model and the known statistics of the measurement noise n_k. Since the measurements of time-step k are used to modify the prior PDF for obtaining the posterior PDF, this step is called *update step*. The complete principle of the recursive Bayesian estimator is presented in Figure 5.2, for example, (Krach et al. 2008).

Finally, the solution that maximizes the posterior PDF is the *MAP estimator*

$$\hat{s}_{\text{MAP},k} = \underset{s_k}{\text{argmax}}\, p\left(s_k|y_1,y_2, \cdots ,y_k\right).$$

Contrary to that, the *minimum mean square error (MMSE)* estimator calculates the expectation of the PDF, that is,

$$\hat{s}_{\text{MMSE},k} = \int s_k\, p\left(s_k|y_1,y_2, \cdots ,y_k\right)ds_k,$$

where for Gaussian noise distributions both estimators yield the same result.

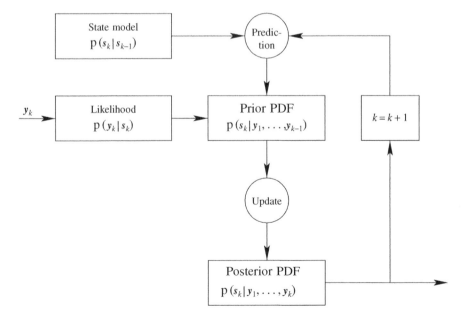

Figure 5.2   Recursive Bayesian estimator

Usually there are no closed-form solutions for the integrals in Bayes estimation (Arulampalam et al. 2002). One option to handle these integrals are certain assumptions for the models or approximations: the Kalman filter approach presumes linear models and Gaussian noise distributions. This is described in Section 5.1. The extended Kalman filter approximates nonlinear models in a linearization step, which will be shown in Section 5.2. Another option to handle the integrals is numerical integration. An approximate solution by Monte Carlo methods is the particle filter described in Section 5.3. All filters will be examined with a special focus on position tracking applications. Finally, in Section 5.4 further advanced tracking approaches are described.

5.1 Kalman Filter

The *Kalman filter (KF)* (see Kalman 1960; Kay 1993) is one of the commonly used implementation of *Bayesian filters*. It is a generalization of the Wiener filter, where the restriction that signal and noise have to be stationary is no longer necessary. It is a *Bayesian sequential MMSE estimator* of a signal embedded in noise, where the signal is characterized by a system model, which is the previously described first order hidden Markov model. One of the main advantages of KFs is the computational efficiency in the implementation using only matrix and vector operations on the mean and covariances of Gaussian processes.

To perform optimally, some assumptions must hold for the system model in Equation (5.2) and the measurement model in Equation (5.1). The *system process noise* $v_k \sim \mathcal{N}(\mathbf{0}_{N_s}, \mathbf{Q}_k)$ and the *measurement noise* $n_k \sim \mathcal{N}(\mathbf{0}_{N_y}, \mathbf{C}_k)$ should be zero-mean Gaussian distributed with given covariances. Moreover, the function $f_{k-1}(s_{k-1}, v_{k-1})$ should be a known *linear function* of s_{k-1} and v_{k-1} and the function $g_k(s_k, n_k)$ should be a known *linear function* of s_k and n_k. If these requirements are met, the KF is the optimum MMSE estimator. If these assumptions do not hold, it is still the optimum linear MMSE estimator.

For the fulfilled assumptions, we can rewrite Equations (5.2) and (5.1) as

$$s_k = A_{k-1}s_{k-1} + v_{k-1}$$

and

$$y_k = H_k s_k + n_k.$$

The matrix $A_k \in \mathbb{R}^{N_s \times N_s}$ is denoted as state matrix and includes the linear dependencies between the states of time-steps k and $k-1$. The measurement matrix $H_k \in \mathbb{R}^{N_y \times N_s}$ reflects the linear relation between the measurements and the state at time-step k. In general, all matrices can be time-variant. In the context of positioning applications this could reflect, for example, changing mobility models over time. The optimum filter equations can then be written as follows.

Since all PDFs (including prior, posterior, and likelihood) are Gaussian, they can be represented by means and covariances. This allows a simple notation of the estimates in terms of matrix-vector notation. In a *first step (prediction)*, the state of the current time-step is calculated taking into account the state of the previous time-step and the knowledge of the state matrix given by A_k. Then, the estimate of the state after prediction is

$$\hat{s}_{k|k-1} = A_{k-1}\hat{s}_{k-1|k-1},$$

with the estimate of the previous time-step $\hat{s}_{k-1|k-1}$. Additionally, the corresponding MMSE or covariance matrix after that prediction step can be calculated as

$$M_{k|k-1} = A_{k-1}M_{k-1|k-1}A_{k-1}^{\mathrm{T}} + Q_k,$$

where $M_{k-1|k-1}$ is the MMSE matrix of the previous time-step. From the Bayesian PDF point of view, the *prior PDF* can be represented as a Gaussian distribution according to

$$p\left(s_k|y_1,y_2, \ldots ,y_{k-1}\right) \sim \mathcal{N}\left(\hat{s}_{k|k-1},M_{k|k-1}\right).$$

The *Kalman gain matrix* includes a weighting between the predicted estimate (already calculated) and the current measurements. It is given as

$$K_k = M_{k|k-1}H_k^{\mathrm{T}}\left(C_k + H_kM_{k|k-1}H_k^{\mathrm{T}}\right)^{-1}.$$

Finally, the correction step combines the predicted estimates with the current measurements weighted with the Kalman gain matrix. This results in the final estimate of the state vector

$$\hat{s}_{k|k} = \hat{s}_{k|k-1} + K_k\left(y_k - H_k\hat{s}_{k|k-1}\right).$$

The corresponding MMSE or covariance matrix after the correction step is obtained as

$$M_{k|k} = \left(I_{N_s} - K_kH_k\right)M_{k|k-1}.$$

The resulting *posterior PDF* can then be written as Gaussian distribution according to

$$p\left(s_k|y_1,y_2, \ldots ,y_k\right) \sim \mathcal{N}\left(\hat{s}_{k|k},M_{k|k}\right).$$

The KF is initialized with $s_{0|0}$ and $M_{0|0}$ determined by the *a priori* distribution of the initial state.

We observe that the MMSE matrix can be calculated independently of the state estimates. Furthermore, it only depends on model parameters and not on actual measurements. Hence, it can be calculated in advance or 'off-line' and provides the expected accuracy for the state estimates over time without processing the measurements.

In the following, we discuss the *KF in the context of position tracking* applications. For this example, we assume that the state vector that has to be estimated consists of position and velocity, that is,

$$s_k = \left[x_k, y_k, v_{x,k}, v_{y,k}\right]^{\mathrm{T}}.$$

For simplification, we restrict the discussion to the two-dimensional case as an extension to the three-dimensional case is straightforward. For the mobility model, we choose a very simple model corresponding to the principle of *random walk*. For that, the resulting system matrix is given as

$$A = \begin{bmatrix} 1 & 0 & T & 0 \\ 0 & 1 & 0 & T \\ 0 & 0 & 1 & 0 \\ 0 & 0 & 0 & 1 \end{bmatrix}$$

The sampling time T highly depends on the application. For instance, for pedestrian navigation sampling times of around 1 s are usually sufficient. The covariance matrix of the process noise is the diagonal matrix

$$Q = \begin{bmatrix} \sigma_{Q,x}^2 & 0 & 0 & \\ 0 & \sigma_{Q,y}^2 & 0 & 0 \\ 0 & 0 & \sigma_{Q,v_x}^2 & 0 \\ 0 & 0 & 0 & \sigma_{Q,v_y}^2 \end{bmatrix}.$$

It includes the variance of the mobility (process drift) in $x-$ and $y-$ direction for the position and velocity. This model implies that the change of the MT position is controlled by process noise of a certain variance. Note that more realistic mobility models will be presented in Chapter 6.

For the measurement model, we assume that in every time-step a position estimate is available. This could be realized by the algorithms described in Chapter 4 for static position estimation. Hence, the measurements are given in terms of position measurements and have a linear dependency on the state vector, which is reflected in the measurement matrix

$$H = \begin{bmatrix} 1 & 0 & 0 & 0 \\ 0 & 1 & 0 & 0 \end{bmatrix}.$$

We do not consider any velocity estimates that are available from the position estimation entity. Thus, the *velocity* is handled as a *hidden state* and estimated implicitly in the filter equations.

Figure 5.3 shows a typical scenario for MT tracking using BSs from a cellular network. The distance between the BSs is set to 1500 m and two generated MT tracks are shown. One at the cell edge between the BSs (track 1) and one close to a BS (track 2). For position estimation TDOA measurements with the nearest three BSs are performed, where it is assumed that the timing estimation standard deviation on each of the links is 50 m.

Figure 5.4 shows the corresponding results in terms of RMSE (see Section 4.5.1) over time for the two tracks and different algorithms. The RMSE was obtained by averaging over several noise realizations. The static solution (as per the reference) is calculated by using the Gauss–Newton algorithm (see Section 4.3). It provides the estimated position (with the corresponding covariance matrix) in each of the time-steps without including any time correlations. For track 1 (cell edge) the geometric situation is reasonably good and the overall position estimation accuracy is around 50–60 m. Contrary to that, for track 2 (close to BS) only at the begin of the track the geometric situation is reasonably good. Here, the position estimation accuracy is around 70 m and 80 m. Getting closer to the BS at the end of the track (from around time-step 160 on ward) the quality of the position estimates becomes worse, which corresponds to the CRLB analysis shown in Section 4.5. In some situations, no accurate position estimate is possible.

Applying the *KF to the static solution*, the overall static position estimates can be 'smoothed' out by taking into account the mobility model. For track 1, one can achieve an accuracy of around 20 m. In the first half of track 2, a similar performance can be obtained. For the second half the inaccurate static position estimates have an impact on the KF estimates too. Nevertheless, these estimates are implicitly weighted by the KF being quite low and more information about the mobility model is used. Despite this, the KF estimates drift more and more over time since not so many useful position estimates are available.

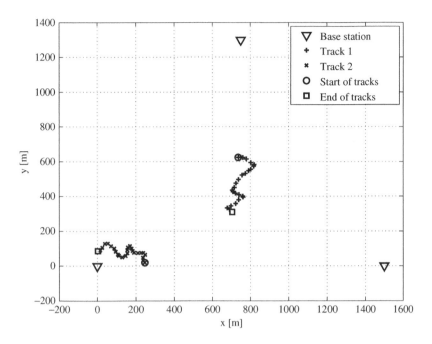

Figure 5.3   Scenario for position tracking

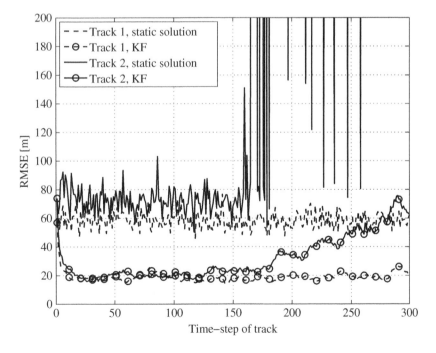

Figure 5.4   RMSE for position tracking with KF

5.2   Extended Kalman Filter

The performance of the KF is optimal if the conditions on Gaussianity and linearity are fulfilled completely. However, already in the simple example shown in Figure 5.4 it becomes obvious that in certain situations (e.g., if the MT is close to a BS) the performance is limited. Furthermore, the KF requires that the underlying entity that provides the static solutions performs optimally, that is, in each time-step enough sources (at least three) have to be available and further, these static position estimates have to be Gaussian distributed. Particularly in critical positioning situations such as urban canyons or indoors, it may quite often be the case that only less than the required number of sources are available for a certain time. Then, the KF would totally fail since the static solution cannot provide any estimates.

The *extended Kalman filter (EKF)*, for example, (Kay 1993; Maybeck 1979), is a much more flexible tool able to handle directly nonlinear models. We assume that the state model is given as

$$s_k = a_{k-1}\left(s_{k-1}\right) + v_{k-1}$$

and the measurement model can be written as

$$y_k = h_k\left(s_k\right) + n_k.$$

The basic idea of the EKF is a *linearization* of $a_{k-1}\left(s_{k-1}\right)$ about the estimate of s_{k-1}. We obtain

$$a_{k-1}\left(s_{k-1}\right) \approx a_{k-1}\left(\hat{s}_{k-1|k-1}\right) + A_{k-1}\left(s_{k-1} - \hat{s}_{k-1|k-1}\right),$$

with the Jacobian matrix

$$A_{k-1} = \left.\frac{\partial a_{k-1}\left(s_{k-1}\right)}{\partial s_{k-1}}\right|_{s_{k-1}=\hat{s}_{k-1|k-1}}.$$

Equivalently, we linearize $h_k\left(s_k\right)$ about the estimate of s_k, that is,

$$h_k\left(s_k\right) \approx h_k\left(\hat{s}_{k|k-1}\right) + H_k\left(s_k - \hat{s}_{k|k-1}\right),$$

with the Jacobian matrix

$$H_k = \left.\frac{\partial h_k\left(s_k\right)}{\partial s_k}\right|_{s_k=\hat{s}_{k|k-1}}.$$

Obviously, the Jacobians have to be re-calculated in every time-step since they depend on the estimates of the previous time-steps. However, the resulting structure of the EKF as pointed out in the following is very similar to the KF solution.

It starts with the *prediction*, where knowledge of the MT movement model is applied to obtain

$$\hat{s}_{k|k-1} = a_k\left(\hat{s}_{k-1|k-1}\right),$$

with the estimate of the previous time-step $\hat{s}_{k-1|k-1}$. Similarly, the corresponding MMSE or covariance matrix after that prediction step is

$$M_{k|k-1} = A_{k-1}M_{k-1|k-1}A_{k-1}^{\mathrm{T}} + Q_k.$$

Due to the linearization step the resulting estimated *prior PDF* in the Bayesian sense is a Gaussian approximation of the true prior PDF. Hence, the estimated prior PDF is given as

$$p\left(s_k | y_1, y_2, \ldots, y_{k-1}\right) \approx \mathcal{N}\left(\hat{s}_{k|k-1}, M_{k|k-1}\right).$$

The Kalman gain matrix can be obtained by

$$K_k = M_{k|k-1} H_k^{\mathrm{T}} \left(C_k + H_k M_{k|k-1} H_k^{\mathrm{T}}\right)^{-1},$$

Finally, the *correction step* combines the predicted estimates with the current measurements weighted with the Kalman gain matrix. This results in the final estimate of the state vector being calculated as

$$\hat{s}_{k|k} = \hat{s}_{k|k-1} + K_k \left(y_k - h\left(\hat{s}_{k|k-1}\right)\right).$$

The corresponding MMSE or covariance matrix after correction is obtained as

$$M_{k|k} = \left(I_{N_s} - K_k H_k\right) M_{k|k-1}.$$

Also the resulting *posterior PDF* is a Gaussian distribution of the true posterior PDF. It is given as

$$p\left(s_k | y_1, y_2, \ldots, y_k\right) \approx \mathcal{N}\left(\hat{s}_{k|k}, M_{k|k}\right).$$

Since the MMSE matrix depends on the Jacobians (we perform a dynamic linearization for the EKF), an 'off-line' calculation similar to the KF is no longer possible. Further, the EKF has no optimality properties where its accuracy depends on the accuracy of the linearization. Nevertheless, the EKF turns out to be a flexible and robust approach widely used for position tracking applications.

Considering the position tracking example similar to the KF (see Section 5.1), the state vector s_k is defined in the same way. Furthermore, we assume the same *state model*, that is, a linearization of the state equation is not necessary. Therefore, the prediction step is the same for EKF and KF. As measurements, we process directly the TDOA measurements–compared to the KF, where the TDOAs were processed beforehand by a static position estimation yielding position estimates. Since TDOAs are nonlinear in terms of position, for the update step a *linearization of the measurement model* is necessary.

The relation between the position and the corresponding TDOAs is determined in the measurement function

$$h_k\left(s_k\right) = \begin{bmatrix} h_{2,1,k}\left(s_k\right) \\ h_{3,1,k}\left(s_k\right) \\ \vdots \\ h_{N_{\mathrm{BS}},1,k}\left(s_k\right) \end{bmatrix},$$

where

$$h_{v,1,k}\left(s_k\right) = r_{v,k}\left(s_k\right) - r_{1,k}\left(s_k\right), \quad v = 2, 3, \ldots, N_{\mathrm{BS}}$$

defines the TDOAs (see Section 4.3) using the distances between MT and the BSs

$$r_{\mu,k}\left(s_k\right) = \sqrt{\left(x_k - x_\mu\right)^2 + \left(y_k - y_\mu\right)^2}, \quad \mu = 1, 2, \ldots, N_{\mathrm{BS}}.$$

Hence, the corresponding measurement vector consists of $N_{BS} - 1$ noisy TDOA measurements (see Section 4.3.2) performed with N_{BS} BSs, that is,

$$\mathbf{y}_k = \left[d_{2,1,k}, d_{3,1,k}, \dots, d_{N_{BS}-1,1,k}\right]^{\mathrm{T}}.$$

For the calculation of the EKF update step, the Jacobian matrix including the derivatives of the measurement model with respect to the state vector has to be available. For the considered example, it is given by

$$\mathbf{H}_k = \begin{bmatrix} \frac{x_k - x_2}{r_{2,k}(\mathbf{s}_k)} - \frac{x_k - x_1}{r_{1,k}(\mathbf{s}_k)} & \frac{y_k - y_2}{r_{2,k}(\mathbf{s}_k)} - \frac{y_k - y_1}{r_{1,k}(\mathbf{s}_k)} \\ \frac{x_k - x_3}{r_{3,k}(\mathbf{s}_k)} - \frac{x_k - x_1}{r_{1,k}(\mathbf{s}_k)} & \frac{y_k - y_3}{r_{3,k}(\mathbf{s}_k)} - \frac{y_k - y_1}{r_{1,k}(\mathbf{s}_k)} \\ \vdots & \vdots \\ \frac{x_k - x_{N_{BS}}}{r_{N_{BS},k}(\mathbf{s}_k)} - \frac{x_k - x_1}{r_{1,k}(\mathbf{s}_k)} & \frac{y_k - y_{N_{BS}}}{r_{N_{BS},k}(\mathbf{s}_k)} - \frac{y_k - y_1}{r_{1,k}(\mathbf{s}_k)} \end{bmatrix}.$$

Compared to the KF example, where an intermediate step of static position estimation was required, the EKF can handle situations with too less measurements flexible. Even if no measurements are available for a certain time, the EKF can provide reasonable estimates. In these situations, only the mobility model is used to update the states. However, the main limiting performance factor in this example is the geometric situation for track 2, where the corresponding results for KF and EKF are plotted in Figure 5.5. We observe, that in the situation at the end of track 2 the EKF can provide more reliable estimates than the KF. Where the KF diverges over

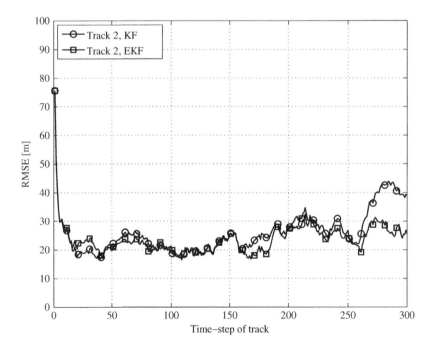

Figure 5.5   RMSE for position tracking with EKF, track 2

time due to a lack of reliable position estimates from the underlying static position estimation, the EKF can follow the true MT track much better.

5.3   Particle Filter

Another important class of Bayesian filters is based on approximation of integrals by numerical integration. These methods are commonly denoted as particle filters (PFs) (Arulampalam et al. 2002; Djuric et al. 2003; Ristic et al. 2004) and became quite popular for position tracking applications, for example, (Gustafsson et al. 2002). PFs are based on a *sequential Monte-Carlo methodology* (see Doucet et al. 2001) and calculate recursively the relevant PDFs by importance sampling and approximation of PDFs with discrete random measures. The basic principle of particle filtering is the representation of the state PDF by a defined number of hypotheses, hence, it does not implement an analytical function. The PF approximates the optimal solution numerically based on the physical system model rather than applying an optimal filter to an approximate model as can be seen for the KF. Compared to KFs, the PFs have usually a much higher complexity depending on the number of particles that have to be generated to model the PDF. In addition, they can suffer from phenomena like *sample degeneracy* or *sample impoverishment* causing unstable behavior.

In PFs, the *posterior PDF* is represented as the weighted sum

$$p\left(s_k | y_1, y_2, \ \ldots \ , y_k\right) = \sum_{i=1}^{N_p} w_k^i \delta \left(s_k - s_k^i\right),$$

where each *particle i* consists of a *state* s_k^i and a *weight* w_k^i, and $\delta\left(\cdot\right)$ is the *Dirac delta measure*. The particles are drawn according to the principle of *importance sampling* from a *proposal density* $q\left(s_k | s_k^i, y_k\right)$. The corresponding weights can then be calculated by

$$w_k^i \propto w_{k-1}^i \ \frac{p\left(y_k | s_k^i\right) p\left(s_k^i | s_{k-1}^i\right)}{q\left(s_k | s_k^i, y_k\right)}.$$

The *generic PF* applies the *optimum proposal density* that in practice is difficult to use. Therefore, often the so-called *sampling importance resampling PF (SIR-PF)* is implemented (see Arulampalam et al. 2002; Gordon et al. 1993; Ristic et al. 2004). It only requires that the state and measurement functions $f_k\left(\cdot\right)$ and $g_k\left(\cdot\right)$ are known, and that sampling of realizations from the state noise distribution of v_{k-1} as well as the prior distribution is possible. In addition, the likelihood function $p\left(s_k | y_k\right)$ has to be available for pointwise evaluation. So compared to the generic PF it can be said that for the SIR-PF the proposal density is chosen to be the prior density according to $p\left(s_k | s_{k-1}^i\right)$.

In the first step of SIR-PF, for each particle $i = 1, 2, \ \ldots \ , N_p$ a sample from the proposal density has to be drawn, that is,

$$s_k^i \sim p\left(s_k | s_{k-1}^i\right).$$

This can be realized by generating a state noise sample v_{k-1}^i with the corresponding PDF $p_v\left(v_{k-1}\right)$ and setting

$$s_k^i = f_{k-1}\left(s_{k-1}^i, v_{k-1}^i\right).$$

In a second step, for each particle the weights have to be calculated. With the chosen proposal density, this step reduces to

$$w^i_k = \text{p}\left(y_k|s^i_k\right).$$

Finally, all weights have to be normalized by

$$w^i_k = \frac{w^i_k}{W},$$

using

$$W = \sum_{i=1}^{N_{\text{p}}} w^i_k.$$

A major problem of the PF is the *degeneracy phenomenon* (Arulampalam et al. 2002; Ristic et al. 2004). It points out that after a few iterations all but one particle will have weights very close to zero. It has been shown that the variance of the weights can only increase over time, and hence, it is not possible to avoid the degeneracy problem. One approach to reduce this effect is simply to use a very large number of particles. However, this is often too inefficient from a computational complexity point of view. A better method is the application of *resampling* where degeneracy can be reduced remarkably. The idea is an elimination of particles with low weights to concentrate on particles with large weights. In this manner, a new set of states $\tilde{s}^i_k$, $i = 1, 2, \ldots, N_{\text{p}}$ is created by resampling N_{p} times from an approximate discrete representation of

$$\text{p}\left(s_k|y_1, y_2, \ldots, y_k\right) \approx \sum_{i=1}^{N_{\text{p}}} w^i_k \delta\left(s_k - s^i_k\right).$$

Given

$$\text{p}\left(\tilde{s}^j_k = s^i_k\right) = w^j_k,$$

the resulting sample is an i.i.d. sample from the discrete density. Even though, the degeneracy can be reduced by resampling, another effect denoted as *sample impoverishment* is introduced in practical implementations. Besides the problem of limited parallelization due to the fact that the particles have to be combined, particles with large weights are statistically selected more often then the other particles. So the diversity among the particles is reduced since the resulting sample will contain many repeated points. Particulary for systems with small state noise, the sample impoverishment can be a serious problem and all particles can be concentrated to a single state after a few iterations.

In addition to the SIR-PF, there exist several other PF approaches in the literature. Briefly, we should mention at this point the *auxiliary SIR-PF (ASIR-PF)* (Arulampalam et al. 2002; Pitt and Shephard 1999). It can be seen as an implementation of the SIR-PF with resampling at the previous time-step, that is, the ASIR-PF generates points from the samples at time-step $k - 1$. They are usually much closer to the true state if conditioned on the current measurements. Compared to the SIR-PF, the ASIR-PF is not so sensitive to outliers if the state noise is small. Further, the weights are distributed more evenly. However, for large state noise the performance of the ASIR-PF can be worse compared to the SIR-PF.

Another PF implementation is the *regularized PF (R-PF)* (Arulampalam et al. 2002; Musso et al. 2001). Compared to the SIR-PF, the R-PF has a different resampling stage. Whereas in the SIR-PF the resampling is done based on a discrete approximation, the R-PF resamples from a continuous approximation of the posterior PDF where a Kernel approach is applied. The R-PF outperforms the SIR-PF particularly in situations where sample impoverishment limits the performance. This could be situations with, for example, low state noise.

Figure 5.6 shows the results for an implementation of the SIR-PF (using $N_p = 1000$ particles) compared to EKF and KF. We observe that only a small performance improvement can be achieved by the PF compared to EKF at the end of track 2. Obviously, the nonlinearities can be approximated well by the EKF. Furthermore, the Gaussian assumptions are fulfilled, and hence, the EKF performs near-optimum already. This can also be observed in Figure 5.7. For the performance criterion we used the CDF of the two-dimensional position error (Section 4.5.2). The CDF was averaged over several MT tracks in the scenario (see Figure 5.3) and noise realizations. It is obvious that a lot of performance can be gained by tracking algorithms in general compared to the static solution. However, in this scenario, as for the well-behaved assumptions like Gaussian processes, the additional gain by an EKF to the KF is quite small. Furthermore, considering the complete scenario, the PF performs nearly the same as the EKF. For instance, in 90% of situations the position estimation error is smaller than 150 m for the static solution. This can be improved to 40 m using KF tracking or 'smoothing'. With an EKF this can be further reduced to 35 m, where the PF performs nearly the same as compared to the EKF.

This becomes different if we use more realistic implementations of the measurements. Figure 5.8 shows various track realizations in an urban canyon scenario. By ray tracing, channel impulse responses were created to reflect multipath as well as NLOS propagation

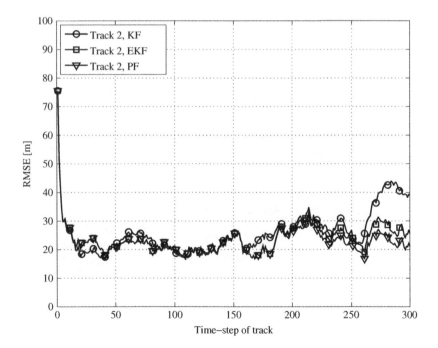

Figure 5.6 RMSE for position tracking with KF, EKF, and PF, track 2

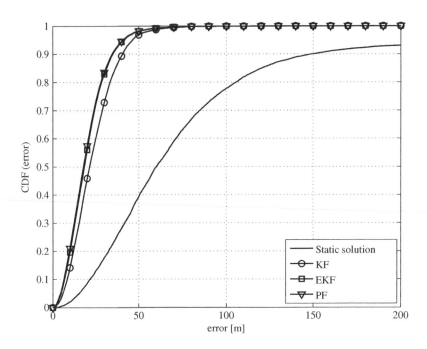

Figure 5.7   CDF for position tracking with KF, EKF, and PF

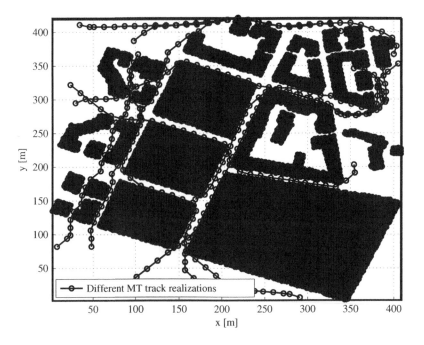

Figure 5.8   Different track realizations in an urban canyon scenario

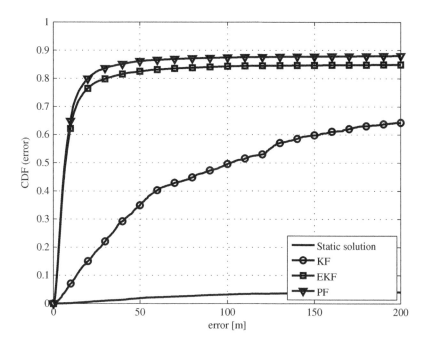

Figure 5.9 CDF for position tracking in an urban canyon scenario

in the simulations, resulting in highly non-Gaussian observation noise. More information on this approach will be presented in Chapter 6. Note that furthermore, the applied mobility model that generates the MT tracks does not completely match the mobility model included in the filters. In Figure 5.9, the corresponding CDFs for position tracking in an urban canyon scenario are depicted. First of all, we observe that positioning with the static solution is not possible in this scenario. Only with tracking algorithms can a sufficient quality of the estimates be achieved. For this scenario, we also can observe distinctive gaps between KF, EKF, and PF.

The CDF curves for an indoor scenario are shown in Figure 5.10. Since static solution and KF do not provide any useful results, only the EKF and PF are plotted. We observe that the position estimation performance for indoor is–as expected–even worse than outdoor. For this example, only in less than 70% of the situations can a position estimate be delivered at all.

5.4 Further Approaches

In the literature, numerous approaches to position tracking or tracking in general are investigated. In the following, we would like to present an overview of more sophisticated approaches based on the general algorithms described before (also see Krach 2010; Mensing 2013).

5.4.1 Grid-Based Methods

If the state space is discrete, and hence, does only include a finite number of states, grid-based approaches can provide the optimum solution (Arulampalam et al. 2002; Ristic et al. 2004).

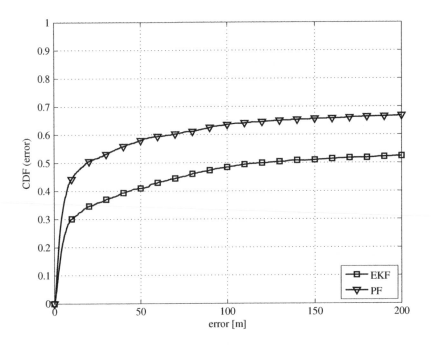

Figure 5.10   CDF for position tracking in an indoor scenario

Assuming that the state vector at time-step $k-1$ consists of N_g states $s_{k-1}^i, i = 1, 2, \ldots, N_g$. Then, the posterior PDF can be represented as a weighted sum of the discrete states, that is,

$$p\left(s_{k-1}|y_1, y_2, \ldots, y_{k-1}\right) = \sum_{i=1}^{N_g} w_{k-1|k-1}^i \delta\left(s_{k-1} - s_{k-1}^i\right). \tag{5.5}$$

Using Equation (5.5) with Equation (5.3), the prediction equation can be written as

$$p\left(s_k|y_1, y_2, \ldots, y_{k-1}\right) = \sum_{i=1}^{N_g} w_{k|k-1}^i \delta\left(s_k - s_k^i\right),$$

with the predicted weights

$$w_{k|k-1}^i = \sum_{j=1}^{N_g} w_{k-1|k-1}^j p\left(s_k^i|s_{k-1}^j\right).$$

Substituting Equation (5.5) into Equation (5.4), the update equation is given by

$$p\left(s_k|y_1, y_2, \ldots, y_k\right) = \sum_{i=1}^{N_g} w_{k|k}^i \delta\left(s_k - s_k^i\right),$$

with the updated weights

$$
w^i_{k|k} = \frac{\sum_{j=1}^{N_g} w^j_{k|k-1} \mathrm{p}\left(y_k|s^j_k\right)}{w^i_{k|k-1}\mathrm{p}\left(y_k|s^i_k\right)}.
$$

5.4.2 Second Order Extended Kalman Filter

The EKF only takes into account a linearization around the current state estimate. To approximate the nonlinearities in a better way, as well as the Jacobian matrices Hessian matrices, including the second derivatives, can be included. These filters are denoted as second order extended Kalman filters or modified Gaussian second order filters. Derivations in the context of position tracking applications can be found in (Ali-Löytty 2009) and (Bar-Shalom et al. 2001). Even more sophisticated approaches including robust extended Kalman filtering are proposed in (Perälä and Piché 2007).

5.4.3 Unscented Kalman Filter

The EKF has two important drawbacks (Wan and van der Merwe 2000). On the one hand, the derivation of the Jacobian matrices, that is, the linear approximations to the nonlinear functions, may be complex and can cause implementation difficulties. On the other hand, these linearizations can lead to filter instabilities if the time-step intervals are not sufficiently small, especially in highly nonlinear environments. To address these limitations the *unscented KF (UKF)* can be an alternative (Julier and Uhlmann 1997). The philosophy of the UKF is that it uses the premise that it is easier to approximate a Gaussian distribution than it is to approximate an arbitrary nonlinear function. Hence, instead of linearizing the system by Jacobian matrices the UKF uses a deterministic sampling approach to capture the mean and covariance estimates with a minimal set of sample points. These points are chosen so that their mean, covariance, and possibly also higher order moments match the corresponding Gaussian random variables.

The idea of the UKF is based on the unscented transform that was proposed by (Uhlmann 1994) with the approach that the nonlinear function is applied to a set of points and the statistics of the transformed points can be used to estimate the transformed mean and covariance. Contrary to PFs, the chosen points are not drawn randomly. These so-called sigma points are chosen deterministically with respect to certain given mean and covariance properties. These sigma points are then propagated through the nonlinearities determined by the state or measurement model. Afterwards, the sigma-points are weighted and finally recombined to produce the estimated mean and covariance. Different approaches to obtaining a valid set of sigma points are provided in (Julier and Uhlmann 2004).

5.4.4 Gaussian Mixture Filter

Gaussian mixture filters or Gaussian sum filters approximate prior and posterior PDFs by Gaussian mixtures, which is a convex combination of Gaussian PDFs. It can be seen as an extension or generalization of the Kalman filtering approach. Limiting factor of Gaussian mixture filters is the number of included PDFs (similar as the number of particles in PF approaches). A good overview in the context of position tracking can be found in (Ali-Löytty 2009).

5.4.5 Rao–Blackwellization

With the increasing number of states that have to be estimated, the number of required particles also needs to be increased. For instance, a state vector in inertial navigation requires a very high number of particles. The idea of Rao–Blackwellization (see Schön et al. 2005), which is sometimes denoted as marginalization, is to reduce the number of particles by using a KF for the part of the system model that is linear. The nonlinear part of the system model is still treated by a PF. Therefore, we split the state vector into a linear and a nonlinear part according to

$$s_k = \left[s_{\mathrm{lin},k}, s_{\mathrm{nl},k} \right]^{\mathrm{T}}.$$

Then, the Rao–Blackwellized PF factorizes the posterior PDF into

$$p\left(s_{\mathrm{lin},k}, s_{\mathrm{nl},k} | y_1, y_2, \ldots, y_k\right) = p\left(s_{\mathrm{lin}} | s_{\mathrm{nl}}, y_1, y_2, \ldots, y_k\right) p\left(s_{\mathrm{nl}} | y_1, y_2, \ldots, y_k\right),$$

where each of the conditional PDFs is handled by different filters. For the example of linear and nonlinear separation, this could be a KF and PF. But this is not a limitation of the Rao–Blackwellized filter since it allows arbitrary combinations of filters (e.g., KF and UKF as analyzed in Briers et al. 2003). It can be shown that, for a Rao–Blackwellized PF with a fixed number of particles, the performance is always better than for a standard PF. Also nested combinations handling more than two different sub-states are possible. For instance, in (Krach 2010; Krach and Weigel 2009) an implementation that combines KF, PF, and grid based methods is investigated.

5.4.6 Map Matching

In the context of particle filtering map matching is often discussed. The idea of map matching is to use map information as additional or side information in the filtering algorithms. For instance, if a street map is available it can be easily included in car tracking algorithms. Since a car is usually only allowed to drive on the streets, hypotheses or estimates being 'off-road' can be dropped out. In the same way, floor plans of buildings can be used to improve pedestrian tracking. In this context, pedestrians cannot go through walls, which must be considered in the algorithms. One straightforward approach can be implemented in a PF. For instance, particles that proceed through walls can be weighted quite low or even dropped out. Also for EKF implementations, map matching solutions have been recently proposed (Perälä and Ali-Löytty 2008).

6

Scenarios and Models

6.1 Scenarios

In this section, we point out *scenarios*, where *positioning with mobile radio systems* can take place. We highlight the typical constraints in these situations and evaluate them qualitatively. Furthermore, we investigate these scenarios in the context of GNSS-based positioning to address limits of satellite navigation and to consider complementary effects. We are especially interested in scenarios where GNSS does not work with the specified accuracy or does not work at all (see GRAMMAR 2009b). In these critical scenarios, the conditions are no longer as desired and a degradation of the performance in terms of accuracy, coverage, and acquisition time is observable depending on the number and quality of the available sources.

Before we start with a description of the scenarios, the fundamental differences between positioning using *GNSS* and *mobile radio systems* should be pointed out:

- *Bandwidth*: As discussed in Chapter 3, the signal bandwidth has a direct impact on the achievable positioning accuracy of the system. GNSS in the mass-market segment targets approaches with a transmit signal bandwidth of around $1-2$ MHz. The desired bandwidths for cellular mobile radio systems like *LTE* are specified up to 20 MHz. For some applications systems with up to 100 MHz are also under discussion, for example, *LTE-Advanced (LTE-A)*. However, here it is still questionable as to whether this spectrum can be allocated as a whole or only realized by aggregation of certain spectrum bands. Nevertheless, the bandwidths in cellular mobile radio systems are much greater compared to GNSS.
- *Sensitivity*: Compared to GNSS, where the distance between satellites and MT is more than 20 000 km, the overall received power level is comparatively very strong for mobile radio systems.
- *Received power level:* Due to the very large distances between satellites and the MT, the received power level is very similar for all received satellites. In cellular mobile radio systems, usually one strong signal from the serving BS is received, whereas other (out-of-cell) BSs are only received with reduced power levels. Furthermore, the trend favors systems with a frequency re-use of one (due to spectrum scarcity) resulting in an unfavorable interference situation.

Positioning in Wireless Communications Systems, First Edition. Stephan Sand, Armin Dammann and Christian Mensing.
© 2014 John Wiley & Sons, Ltd. Published 2014 by John Wiley & Sons, Ltd.

- *Signal design*: As GNSS is a designated positioning system, the signals (or at least most part of the signals) are *a priori* known at the MT. The communications part of GNSS is restricted to a low data rate signal which is modulated 'on-top' of the navigation part and can be easily extracted. Mobile radio systems are designed for communications. Therefore, the amount of *a priori* known sequences is limited to the absolutely necessary (e.g., synchronization sequences, pilots for channel estimation). Moreover, these sequences are not properly designed for positioning purposes, that is, they are not optimized for high precise positioning as it was done for GNSS.

- *Network synchronization*: The satellites in GNSS use atomic clocks and ground-based augmentation for precise network synchronization in time and frequency. For the desired applications, they can be assumed to be perfectly synchronized. In cellular mobile radio systems, it is not *a priori* given that the network is synchronized at all. It might only be an optional feature in certain standards. Additionally, in network synchronization for communications, time synchronization on frame or symbol level might be sufficient, which would result, for example, in positioning errors of several kilometers in LTE. Here, more sophisticated approaches (like network synchronization using GNSS or implemented in terms of location measurement units) are necessary to enable the system for precise positioning.

- *Position of sources*: In GNSS, the precisely known satellite positions are part of the signal which can be exploited by the MT. On the other hand, the positions of the BSs (and specifically of the antennas) are not necessarily known with the desired accuracy.

- *Propagation conditions*: GNSS assumes usually LOS conditions between satellites and the MT. It will be visualized later in this section, that under optimum free space situations a reasonable number of satellites are available, such that at least four of them can be accessed LOS. However, particularly in urban canyons or indoor situations this is no longer the case. In mobile radio systems, the LOS case, especially to more than one BS, is usually the exception. They operate mostly under critical NLOS conditions.

- *Position processing*: GNSS is inherently designed as a system for three-dimensional positioning. In mobile radio systems the overall geometric constellation makes it difficult to extract height information as BSs and MT are normally at a similar height.

6.1.1 Rural Environment

The rural scenario is a wide area scenario. From a GNSS point of view the signals are mainly disturbed by reflections and attenuation. For reflections, the main multipath effects may come from mountains and hills, but also from trees and groups of trees like alleys and forests. However, due to the large distant multipath effects (relevant are only additional paths with an equivalent path length of smaller than about 300 m for GPS, depending on correlator spacing and chip length), these are an assessable problem. Another reason for critical effects is foliage that causes a strong decrease in the received signal strength. Also the structure and properties of the terrain, for example, water or grass, have an impact on the expected performance. Nevertheless, usually comparable very good LOS-access to the satellites is given for this scenario. Moreover, the geometric constellation of the satellites gives a reasonably good geometric condition resulting in an overall good accuracy of position estimates in all three dimensions.

From a mobile radio positioning point of view, in rural areas the cell sizes can be very large, for example, cell radii up to 15 km are commonly used in GSM, for LTE cell radii up to 100 km are specified. Thus, the MT has access only to a very limited number of BSs, mostly only one. Therefore, positioning algorithms exploiting several BSs cannot be used in a proper way. Hence, in rural environments only rough position information might be available, for example, in terms of cell identities.

6.1.2   Urban Environment

An urban area is an area with a high density of buildings. If streets are cutting through dense blocks of buildings, we can talk about urban canyons. In such a scenario usually only a limited number of satellites can be accessed LOS and the overall satellite access is usually strongly disturbed by *multipath* and *NLOS* effects.

Hence, additional information from the cellular network is required for an improvement of the sensitivity or in terms of supplementary measurements for hybrid processing purposes. The cell size is usually much smaller, for example, a few hundred meters for classical BSs, or even only a few meters for hotspots, compared to rural environments; therefore the access to at least a few BSs is likely. Nevertheless, the cellular network-based estimates, similar to the GNSS measurements, are usually subject to multipath and especially NLOS effects.

The quality for positioning with GNSS is strongly correlated with the *visible part of the sky*, that is, how good LOS access to the satellites is possible. Therefore, we calculate the *aperture angle* of an 'virtual' urban canyon in a three-dimensional scenario. A two-dimensional section is shown in Figure 6.1. In this example, H denotes the height of the buildings and W the width of the street. The MT is assumed to be at a distance of W_{MT} from the left building and is located

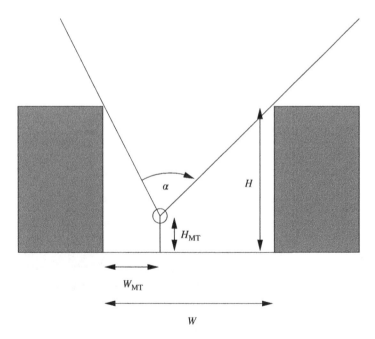

Figure 6.1   Aperture angle in urban canyon scenario

at a height of H_{MT}. Based on this information, we can simply calculate the resulting aperture angle, that is, we can evaluate the percentage of sky that can be seen by the MT, and thus, we have an idea how many satellites can be accessed via LOS for a specific satellite constellation, a specific position, or as an average. We further assume that the urban canyon is of infinite length. With this information, the aperture angle can be computed as

$$\alpha = \arctan \frac{W_{MT}}{H - H_{MT}} + \arctan \frac{W - W_{MT}}{H - H_{MT}}.$$

Figure 6.2 shows the aperture angle α for different building heights H and different distances between the buildings W assuming that the MT is located at $W_{MT} = W/2$ and $H_{MT} = 1.5$ m. We observe, that for $W \sim H$, which is a common situation for urban canyons, the aperture angle is around $50°$. Furthermore, it can be seen that if the building height is much bigger than the width of the street, like in cities with skyscrapers, the aperture angle reduces to 10–$20°$.

Figure 6.3 analyzes the influence of the MT position on the street between the buildings W_{MT} for $W = 20$ m and different building heights H. Clearly, if the MT is located very close to a building the overall aperture angle can be much smaller compared to the situation, where the MT is located directly in the middle between the two buildings. For instance, if the height of the buildings is $H = 20$ m the aperture angle is $\alpha = 57°$ for $W_{MT} = 10$ m. We further observe that the dependency on the MT location is higher for smaller buildings, where for $H = 10$ m the aperture angle varies between 67 and $99°$. On the other hand, for a building heights of $H = 30$ m the variation is only between 35 and $39°$.

These results give an qualitative idea how much of the sky is visible from the MT position, and hence, how many satellites can be accessed LOS. To assess these effects also quantitatively,

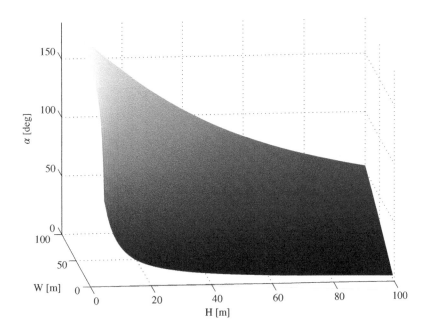

Figure 6.2   Aperture angle α in an urban canyon, width of the street W, height of the buildings H

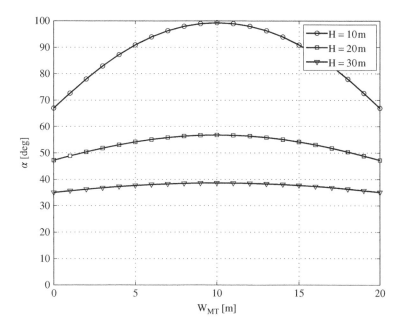

Figure 6.3 Aperture angle α in an urban canyon, width of the street $W = 20$ m, different MT positions W_{MT} and different heights of the buildings H

more information about the position on Earth and the current satellite constellation is needed. In the following, we assume that the MT is located at latitude 48°08′ N and longitude 11°34′ E corresponding to Munich, Germany. The date and time for obtaining the satellite constellation is 2004–01–01, 00:00:00, (UTC). Note that the repetition period for GPS is 23 h 55 min 56.6 s, that is, after that period each satellite is at the same position in the orbit. For Galileo this period is around 10 days.

Figure 6.4 shows a so-called *skyplot* of the constellation for the given coordinates. The skyplot visualizes the situation in the orbit for a given location on Earth and the corresponding date and time. Thus, we can directly observe the positions of the plotted GPS and Galileo satellites in terms of azimuths and elevations. In this example, nine GPS satellites and 10 Galileo satellites could be detected. Also visualized in this skyplot are the orbits of the satellites (thin gray lines). We further observe that there is a 'hole' in the north, that is, an area in the sky, where at no time satellites will be. This is typically for navigation systems and is caused by the constellation of the orbits.

To see the effects of an urban canyon in this example, the plot further includes the border of the aperture angles for an urban canyon with $W = H = 20$ m, that is only the satellites inside the area bounded by the '+' symbols can be accessed LOS, that is, four in our example. Moreover, it can be seen that the orientation of the channel has an impact on the number of detectable satellites.

As the skyplot considers only the situation for one specific time instance, in the following we average the visibility of the satellites over whole GPS/Galileo periods. Figure 6.5 shows the corresponding results in terms of the CDF for free space and the urban canyon. For free space it is assumed that there are no buildings present in the scenario. It can be seen that in all

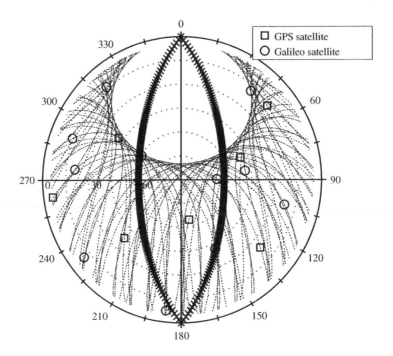

Figure 6.4   Skyplot with limited satellite visibility in an urban canyon

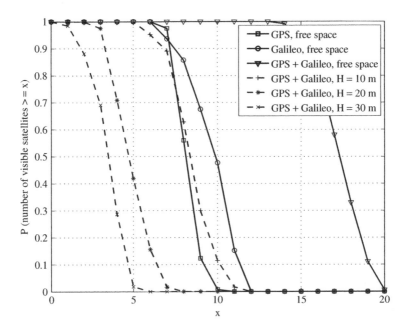

Figure 6.5   CDF for number of visible satellites in an urban canyon

situations at least six GPS or Galileo satellites are visible. On the other hand, this means that always more than the required four satellites are available for positioning. Clearly, this situation changes if we are inside an urban canyon. In this example, the distance between the buildings is $W = 20$ m and the MT is located in the middle of the street. We observe that for $H = 20$ m in 30% of the cases only less than four satellites are visible, and hence, only limited positioning accuracy can be expected. If we increase the building height to 30 m in 70% of cases we have a GNSS-critical situation. For a building height of 10 m more than five satellites can be always accessed LOS. However, here multipath effects will also limit the performance.

Obviously, the geometric constellation of the satellites with respect to the MT position also depends on the properties of the urban canyon. The *GDOP* (see Section 4.5) is a good value to quantify the effect of this impact to the overall positioning accuracy. It determines the dilution factor of the single link accuracy under the respective geometric constellation. Figure 6.6 depicts the CDF for GDOP in free space and the urban canyon. In the free space situation the GDOP is always below four for GPS or Galileo and below two for combined processing of GPS and Galileo. Here, the impact of the geometric effects is rather low even for disadvantageous constellations. If we are in an urban canyon, for example, $W = 20$ m, MT located in the middle of the street, it can happen quite often that the visible satellites are on a 'line' directly above the MT as it can be imagined already in Figure 6.4. Here, just the geometric effects limit the positioning capabilities drastically. Even if at least four satellites are available, a GDOP > 10 makes precise positioning impossible. In these situations a fusion with mobile radio based systems is especially useful.

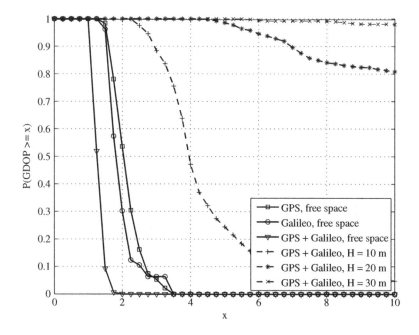

Figure 6.6 CDF for GDOP in an urban canyon

6.1.3   Transition from Outdoor to Indoor

The transition from outdoor to indoor is a scenario with low MT mobility. The SNR loss can be dramatical in such an environment, and in certain cases no satellites can be acquired or not even tracked. In this limiting case only a pure wireless communications based navigation is possible. Due to the blocking of the LOS component and multipath propagation causing amplitude fading for more and more GNSS signals, the SNR loss can be critical in such an environment and more and more satellites will drop out.

6.1.4   Indoor Environment

In pure indoor environments the SNR loss can exceed in certain cases the acquisition and tracking thresholds of GNSS receivers, that is, no satellite can be acquired or tracked. In this limiting case only a pure mobile radio based navigation is possible or the cellular system supports in terms of *assisted GNSS* to track at least a few satellites. For visualization, Figure 6.7 shows the skyplot from our example previously, where the sky view is limited in an indoor environment. Here it is assumed that the MT is located in the middle of a room with dimensions $10 \times 8 \times 4$ m. Only the area of this 'box' directing to the East direction is open. We observe

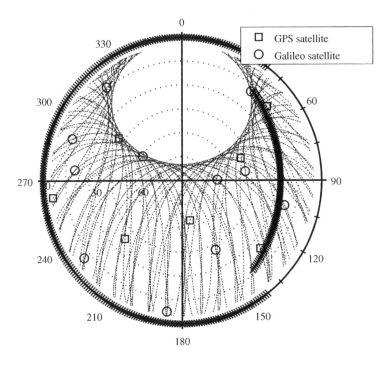

Figure 6.7   Skyplot with limited satellite visibility in an indoor environment

that for this example only two satellites can be accessed directly. However, in practice also attenuation of the windows will further limit the reception.

6.2   Channel Characterization

6.2.1   Channel Measurements

The current channel models for positioning applications are mostly not accurate enough, that is, they are not fine enough in resolution for the investigation of precise localization techniques. Therefore, high accurate channel measurements are needed to analyze the channel properties accurately, for example (Wang et al. 2009a). Clearly, there are numerous channel models available for communications purposes. Nevertheless, they usually lack of important parameters that are essential for positioning applications. Especially the NLOS bias is not included in state-of-the-art communications models, that is, the additional bias introduced in NLOS environments with respect to the *geometric LOS (GLOS)* path is usually not modeled. This information is not needed for investigating communications systems. However, it is fundamental for modeling positioning systems. Another example is the multi-link capability of the channel models. For positioning, measurements with at least three BSs are required that should be modeled coherently. The output of a channel measurement campaign is usually a set of *channel impulse responses (CIRs)* over time.

6.2.2   Ray Tracing

Channel measurements are usually very expensive in terms of the required equipment. Furthermore, their planing and execution takes a lot of time and resources. An alternative for quick assessment of channel properties, also over a wider range, are ray tracing approaches, for example, (WHERE D3.1 2009). Particularly, when different systems, for example, GNSS and wireless communications systems, should be investigated at the same time realistic channel models for these systems are also required, where strong system, spatial, and time correlations occur. Realistic models by using ray tracing simulations in certain scenarios can generate the relevant CIRs for, for example, satellites and BSs at several points. This allows a coherent simulation for the algorithms of interest. One randomly chosen path track realization of a scenario located in Munich, Germany, is depicted in Figure 6.8. Typical CIRs for such a track assuming a BS and a GPS satellite are depicted in Figures 6.9 and 6.10. They were obtained by ray tracing simulations. We observe, for example, the high number of reflections and also the clear LOS situations at time-steps 30–150 and 300–350 for the GPS satellite.

6.3   Channel Models

Once channel measurement or ray tracing is performed, the obtained or estimated CIRs can be used directly to investigate positioning algorithms. As a generalization, usually channel models are developed. They have a statistical character and can be conditioned to different parameters, for example, distance, LOS or NLOS status, environment. Since there is a dissimilar focus on

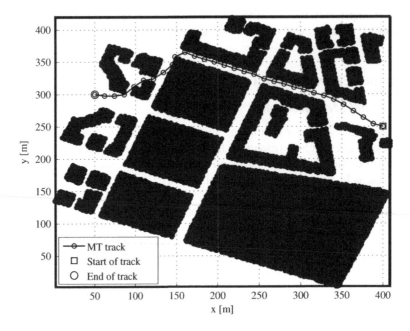

Figure 6.8 Path track realization in urban canyon scenario

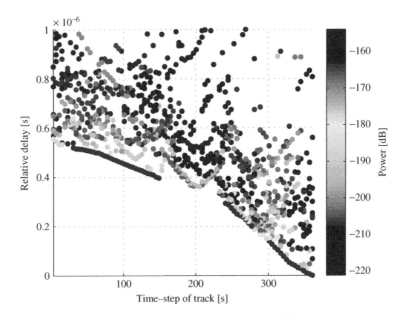

Figure 6.9 Channel impulse response for a GPS satellite

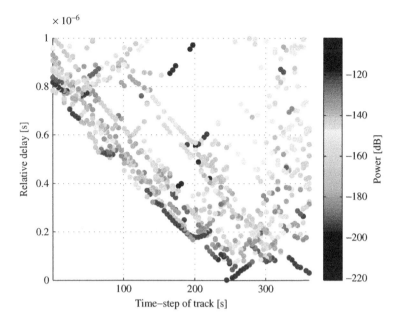

Figure 6.10   Channel impulse response for a BS

positioning and communications, the channel models for both applications are quite different from each other, for example, (Wang et al. 2009b).

6.4   Mobility Models

Tracking algorithms require information about possible movements of the user. This is usually implemented in terms of suitable mobility models restricting the movement within certain constraints. For instance, considering a pedestrian user or a moving car, certain information about the position can be derived using the history of past estimates and suitable movement or mobility models. This includes restricted movements of the user. For instance, a pedestrian cannot 'jump' from one position to another in limited time and, a car can usually only change its direction 'smoothly'. This behavior can then be used as side-information in the position tracking algorithms to improve the performance. Furthermore, mobility models can be used to generate realistic user tracks for simulations and performance evaluation. For notational convenience, we will employ a simple one-dimensional notation for the mobility models in the sequel.

A good and comprehensive overview of *state-space mobility models* can be found in (WHERE D4.3 2008) and (Ali-Löytty 2009). The approach is to define a state vector that includes all the relevant quantities associated with the motion of the target, and then to define its evolution in time by a set of difference equations for the relevant discrete time models. It assumes a general discrete time model that is represented by a first order vector difference equation of the form

$$\boldsymbol{x}_k = \boldsymbol{f}_k(\boldsymbol{x}_{k-1}, \boldsymbol{w}_k),$$

where x_k is the target state and w_k is the process noise vector at the k-th time-step. Here, f_k is a vector-valued function that is often linear and time constant, such that this equation can be rewritten in the linear model

$$x_k = Fx_{k-1} + Gw_k,$$

where F and G are appropriate matrices.

In the *constant velocity model*, also known as *random force model*, the state vector for each spatial dimension is defined as

$$x_k = [x_k, v_k]^T,$$

where x_k and v_k are the MT position and velocity at time step k. Hence, the model can be written as

$$x_k = Fx_{k-1} + Gw_k$$

using

$$F = \begin{bmatrix} 1 & T \\ 0 & 1 \end{bmatrix}$$

and

$$G = \begin{bmatrix} T^2/2 \\ T \end{bmatrix},$$

where T denotes the sampling period and the process noise is white noise with a given standard deviation. This model is often used because of its simplicity. Nevertheless, it is inadequate to emulate users with high mobility (rapidly changing directions or velocities).

In the *constant acceleration model*, the state vector is extended to also include the MT acceleration a_k as parameter, that is,

$$x_k = [x_k, v_k, a_k]^T.$$

The state equation is analogous as before using the matrices

$$F = \begin{bmatrix} 1 & T & T^2/2 \\ 0 & 1 & T \\ 0 & 0 & 1 \end{bmatrix}$$

and

$$G = \begin{bmatrix} T^2/2 \\ T \\ 1 \end{bmatrix},$$

and the process noise is white noise with a given standard deviation. Hence, the mobile acceleration is modeled as a random sequence with independent increments. This model allows a better description of users who change their direction or velocity more rapidly compared to the constant velocity model.

The models described before are rather simple to implement and allow a quick performance evaluation in the context of tracking algorithms. However, they do not reflect any restrictions from the environment like buildings or walls. For that, more sophisticated mobility models are necessary. One example is a model based on a *gas propagation model* to generate a path from a source point to a destination point (see Kammann et al. 2003). Given the source and destination points, building the path is done in two steps: the first step is to compute a gas map that specifies the gas potential at every point in the map. The second step is to obtain the path from the gas map. The gas diffusion algorithm is well-suited to describing human paths in restricted environments. The implementation idea is to simulate gas diffusing into the environment and to obtain a path from these gas diffusion values. This is done by using different matrices that represent different parameters for every point in the map:

- The layout matrix or map: This is simply the map of the environment that holds a 0 for a wall and a 1 for a valid point.
- The gas diffusion matrix or gas map: This matrix holds the gas potential value for every point in the map.
- The filter matrix: This matrix is used as weighting tool in the computation of gas potential values.

The path computation is then done in two steps:

- Fill the gas map by computing the gas potential for every point in the map.
- Use backtracking in the gas map in order to determine the path from source to destination.

For filling the gas map the following principles are used:

- The source point has a gas potential of one, which is unique.
- Gas will spread from the source point to every other valid point on the map, the filling of the gas map ends when all valid points have been computed.
- Gas potential values must be such that if we start at a random point in the map and move in the direction of increasing values, we will eventually reach the source (this is the backtracking used to obtain the path).

Computing the gas value of a point is done by determining the maximum and minimum gas value of the computed neighbors of the current point, computing a weighted average of both values and multiplying by a percentage loss factor according to

$$\text{GasPotential} = (1 - \text{StepLoss})(100\,\text{Gmax} + \text{Gmin})/101.$$

The parameter StepLoss is chosen to be 0.1% and is incorporated to make sure that gas values decrease as soon as it is moved away from the source (see Kammann et al. 2003). The weighted average coefficients are chosen in order to further control the decrease in gas values. The coefficient for Gmin must not be 0 in order to incorporate a wall effect. Walls have gas values of 0 thus lowering the gas values of nearby points. Both parameters control the amount of loss as it is moved away from the source. The loss, however, is always there as long as StepLoss and the coefficients are greater than zero. Hence, the gas map will always be valid for path computation. An example for the gas distribution can be seen in Figure 6.11.

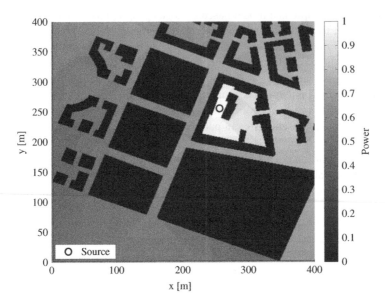

Figure 6.11   Gas distribution for mobility model based on gas diffusion

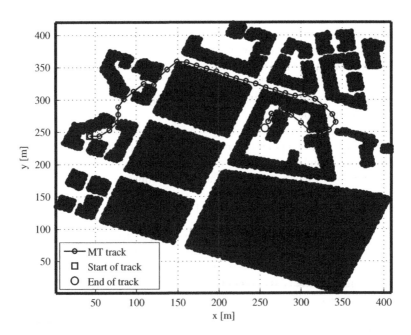

Figure 6.12   Path track generated with mobility model based on gas diffusion

The backtracking and path generation algorithm then uses the created gas map:

- Start at the destination.
- Determine the neighbors of the current position.
- Determine those neighbors with the maximum gas value.
- Randomly select a point between the neighbors with maximum gas value.
- Move to the selected neighbor.
- Repeat until the current position is the source.

It is now obvious why the gas values need to decrease when moving away from the source. If this condition is not met, the backtracking algorithm might loop infinitely (see Kammann et al. 2003). A random generated path track of this method is shown in Figure 6.12.

7

Advanced Positioning Algorithms

The previous chapters described fundamental principles and methods of wireless positioning, so far the focus has been on the position estimation for one single MT using a single source of information and optimum propagation conditions. This chapter will highlight some advanced positioning algorithms in the context of wireless communications systems. We start with the approach of data fusion of location information from multiple sources in Section 7.1. An extension of this method is elaborated on in Section 7.2, where the position of several MTs is determined in a cooperative manner. Finally, we discuss the effects NLOS and multipath propagation on the positioning estimates and elaborate on methods for suitable mitigation (Section 7.3).

7.1 Hybrid Data Fusion

7.1.1 General Hybrid Data Fusion Aspects

The fusion of positioning sensor information is a useful approach to dealing with performance limiting effects. If appropriately designed, every kind and amount of information will increase the accuracy and reliability of the position estimate for the MT. Moreover, sensors that are in principle not suitable for providing stand-alone position information (e.g., compasses or accelerometers), can contribute their individual part of information to improve the overall estimate.

7.1.2 Extension of Derived Algorithms to More Sources

In the following, we will demonstrate, how a cellular communications system can support *WiFi* and *GNSS* in positioning critical environments. To do so, we make use of simplified measurement models for the incorporated systems (see WHERE D2.1 2008).

For the *WiFi system* (see Section 8.4.2), we emulate *RSS* measurements between the MT and the WiFi *hotspots (HSs)*. We assume that the sources, that is, the hotspots, are located at

$$x_\mu^{(HS)} = \left[x_\mu^{(HS)}\ y_\mu^{(HS)}\ z_\mu^{(HS)}\right]^T, \quad \mu = 1, 2, \dots, N_{HS},$$

Positioning in Wireless Communications Systems, First Edition. Stephan Sand, Armin Dammann and Christian Mensing.
© 2014 John Wiley & Sons, Ltd. Published 2014 by John Wiley & Sons, Ltd.

and the MT is located at

$$x = \begin{bmatrix} x & y & z \end{bmatrix}^{\mathrm{T}}.$$

The distance between hotspot μ and the MT is then given by

$$r_\mu^{(\mathrm{HS})}(x) = \sqrt{(x_\mu^{(\mathrm{HS})} - x)^2 + (y_\mu^{(\mathrm{HS})} - y)^2 + (z_\mu^{(\mathrm{HS})} - z)^2}.$$

As a simple approximation and model, we assume that the RSS measurements can be modeled by

$$r_\mu^{(\mathrm{HS})} = r_\mu^{(\mathrm{HS})}(x) + n_\mu^{(\mathrm{HS})},$$

that is, we assume independent *AWGN* components $n_\mu^{(\mathrm{HS})}$ in each link for each included WiFi hotspot. We further assume that the noise contribution in each link is independent from all the other links. Hence, we obtain a covariance matrix of the measurement errors in diagonal form

$$C^{(\mathrm{HS})} = \begin{bmatrix} \sigma_{n_1}^{2,(\mathrm{HS})} & 0 & \cdots & 0 \\ 0 & \sigma_{n_2}^{2,(\mathrm{HS})} & \cdots & 0 \\ \vdots & \vdots & \ddots & \vdots \\ 0 & 0 & \cdots & \sigma_{n_{N_{\mathrm{HS}}}}^{2,(\mathrm{HS})} \end{bmatrix}.$$

Furthermore, we assume that a WiFi hotspot is only visible for the MT if it is in its coverage range.

For the *communications system*, we assume a classical cellular network with a certain inter-site distance. At the MT, the arrival time of signals from several BSs is detected and the difference with respect to a reference BS is used to handle the receiver synchronization time offset with respect to the synchronized BSs, that is, we are using the TDOA approach. The simulated TDOA measurement errors are certainly distance dependent with respect to BS positions and are biased. However, to get a simple model we assume that the TDOAs could be modeled by

$$d_{v,1}^{(\mathrm{BS})}(x) = r_v^{(\mathrm{BS})}(x) - r_1^{(\mathrm{BS})}(x) + n_v^{(\mathrm{BS})} - n_1^{(\mathrm{BS})}, \quad v \in \{2, 3, \dots, N_{\mathrm{BS}}\},$$

that is, we assume independent AWGN in each link for each included BS. The resulting covariance matrix for the TDOA measurements is therefore a non-diagonal matrix given by

$$C^{(\mathrm{BS})} = \begin{bmatrix} \sigma_{n_1}^{2,(\mathrm{BS})} + \sigma_{n_2}^{2,(\mathrm{BS})} & \sigma_{n_1}^{2,(\mathrm{BS})} & \cdots & \sigma_{n_1}^{2,(\mathrm{BS})} \\ \sigma_{n_1}^{2,(\mathrm{BS})} & \sigma_{n_1}^{2,(\mathrm{BS})} + \sigma_{n_3}^{2,(\mathrm{BS})} & \cdots & \sigma_{n_1}^{2,(\mathrm{BS})} \\ \vdots & \vdots & \ddots & \vdots \\ \sigma_{n_1}^{2,(\mathrm{BS})} & \sigma_{n_1}^{2,(\mathrm{BS})} & \cdots & \sigma_{n_1}^{2,(\mathrm{BS})} + \sigma_{n_{N_{\mathrm{BS}}}}^{2,(\mathrm{BS})} \end{bmatrix},$$

reflecting the difference operation with respect to the reference BS.

For the GNSS system, we assume that signals from N_{GPS} *GPS* satellites and N_{Galileo} *Galileo* satellites are acquired and tracked, and hence, $N_{\mathrm{GNSS}} = N_{\mathrm{GPS}} + N_{\mathrm{Galileo}}$ satellites are available for positioning from the GNSS side. For modeling the measurements, we make use of the fact that GNSS positioning can be considered as TDOA based positioning and simply use the same model as for the communications system, however, with different standard deviations of the measurement errors.

If all measurements are available at a certain time-step, the fusion process can start. We assume that the measurement rate is constant for the individual systems and the measurements are available as time aligned. If this is not the case, some measurements might be outdated at the fusion step. This could be taken into account by appropriate weighting. We further assume, that the measurement errors of the different systems are uncorrelated. Depending on the chosen approach or algorithm, for example, static position estimation with the Gauss–Newton method or position tracking with the EKF, the respective vectors and matrices of the individual measurements can simply be combined. As an example, we later will apply the EKF derived in Section 5.2 to this method. Since the mobility model is the same as it is independent of the fusion of several sources, there will be no changes to that. The measurement model will certainly change as the number of available sources is different now. Note that the number of sources can also change over time (for instance if the MT leaves the coverage radius of some sensors).

7.1.3   Simulation Results

In the following we present some simulation results for hybrid data fusion with the shown models, where we start with a *static data fusion* of WiFi RSS measurements and TDOAs from a communications system. The WiFi hotspots are randomly generated in the environment with an average density of 200 hotspots/km^2 and a random coverage radius between 20 and 50 m. The available measurements are impaired by an error according to the fixed standard deviation of $\sigma_n^{(HS)} = 10$ m. For the TDOAs, we assume that measurements with the nearest three BSs are performed in a cellular communications network with an inter-site distance of 1500 m and the measurement error is generated according to the standard deviation $\sigma_n^{(BS)}$, where this parameter is assumed to be 10, 50, and 100 m. The position estimate is done for a static MT using WiFi hotspots, cellular BSs, or a combination of both. Considering the *WiFi-only positioning*, it is usually required that the MT is in the coverage area of at least three hotspots. However, in cases where fewer hotspots are visible to the MT, the following procedure is applied:

- No WiFi hotspot visible: no position solution can be provided.
- One WiFi hotspot visible: the estimate of the MT position is the position of the hotspot.
- Two WiFi hotspots visible: here, the two intersections of the two respective circles are determined. The estimate of the MT position is one of them (randomly chosen).

This limits the position error to the WiFi hotspot coverage area when at least one hotspot is visible. Figure 7.1 shows the simulation results. We observe that in around 24% of the situations no position estimate can be provided for WiFi-only processing. However, when the MT is in the coverage area of a WiFi hotspot, the position estimates are better than 50 m, which reflects the maximum coverage range of the hotspot. The wireless communications based TDOA measurements are also shown in these plots. The 90% error is at around 23 m for a standard deviation of 10 m, at around 110 m for a standard deviation of 50 m, and at more than 200 m for a standard deviation of 100 m. However, the cellular network can provide a much better global coverage, and hence availability, than the WiFi hotspots. Therefore, the hybrid solution that combines WiFi and cellular approaches can provide both reliability and availability. Particularly at high standard deviations for the TDOA measurements, the additional WiFi hotspots can improve the performance.

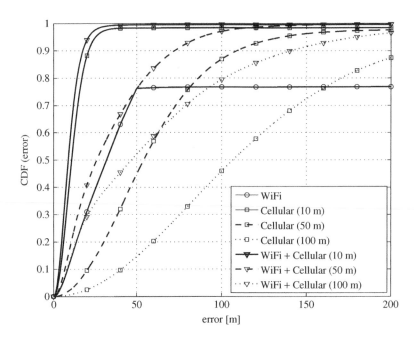

Figure 7.1 Hybrid data fusion of WiFi and cellular communications system

Figure 7.2 shows simulation results for positioning by exploiting hybrid data fusion of a communications system and GNSS. We further assume position tracking of a dynamic MT by applying EKF. In the so-called free space situation access to all potential GNSS satellites is possible. To simulate a more critical environment for GNSS, we further emulate an urban canyon situation. To do so, we assume that not all satellites are visible during the MT track and apply a time dependent satellite visibility, that is, we change the number of visible satellites every 10 s according to the following pattern: All visible, 4, 3, 2, 1, 0, 1, 2, 3, 4, all, 4, 3, and so on. At the begin of the MT track, all satellites are visible. When the MT enters the urban canyon, more and more satellites are dropped out, for example, due to blocking of the LOS signal by buildings, until for a period of 10 s no satellite is visible. Then, more and more satellites can be detected again. We further assume that only the satellites with the highest elevations can be detected. This reflects the reality considering a typical urban canyon situation. For the GNSS measurements, an elevation dependent standard deviation was applied (see ESA 2007), where over several GNSS constellations was averaged. We observe in Figure 7.2 that depending on the quality of the TDOA estimates the average overall positioning error can be reduced remarkably, especially in the critical situations. For instance, the 90% error can be reduced from 27 m (GPS + Galileo) to around 22.5 m (standard deviation of 100 m), 17 m (standard deviation of 50 m), and 7.5 m (standard deviation of 10 m). This then comes really close to the optimum free space situation.

Figure 7.3 shows the final simulation results for a more realistic modeling of the cellular communications system (see WHERE D2.3 2010). To do so, an *LTE system* was simulated on the physical layer level and timing estimation with the three nearest BSs was performed by exploiting LTE's secondary synchronization channel. Moreover, advanced interference cancelation

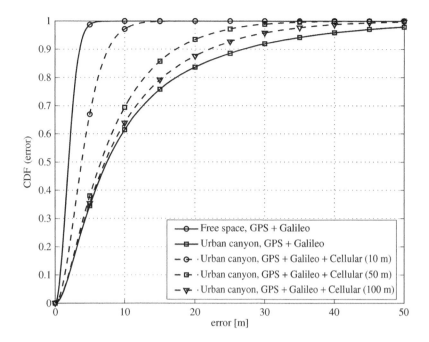

Figure 7.2 Hybrid data fusion of GNSS and cellular communications system, tracking with EKF

and data-aided approaches were applied to increase the performance of the timing estimates. For stand-alone LTE positioning, the 90% error is at around 45 m. Furthermore, the availability of position information within the LTE network is at a very high level. If we fuse these measurements with GNSS, an 90% error of below 20 m can be achieved.

7.2 Cooperative Positioning

7.2.1 General Cooperative Positioning Aspects

As already discussed, for two-dimensional positioning it is required that the MT performs measurements with at least three BSs. If links are blocked, for example by walls in urban canyon environments, or the geometric conditions are restricted, the MT might not be able to determine its position accurately. For such situations, in addition to the previously described hybrid data fusion methods, a cooperative approach can be recommended where MTs can communicate via *peer-to-peer (P2P)* links with each other. On the one hand, that allows the direct exchange of position information between neighboring MTs. On the other hand, these P2P links can be used to derive distance information between these MTs, which can be further exploited for position estimation. This *cooperative positioning (CP)* approach helps to improve the performance in terms of accuracy and coverage compared to conventional hybrid data fusion techniques (see Mensing and Nielsen 2010; WHERE D2.4 2010).

The concept of cooperative positioning, mostly applied nowadays to *wireless sensor networks (WSNs)*, has been recently introduced to heterogeneous communications systems.

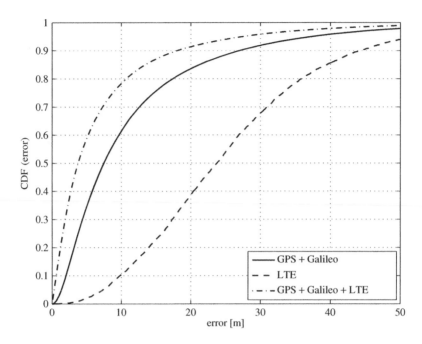

Figure 7.3   Hybrid data fusion of GNSS and LTE, urban canyon, tracking with EKF

However, techniques proposed for WSN cannot straightforwardly be extended to mobile communications networks. This is because these networks usually operate in a very complex wireless environment due to factors like shadowing, mobility, communications infrastructure, or multiple air-interfaces. Hence, the heterogeneity of today's wireless communications networks can be seen as an additional problem in this context.

In principle, there are two different procedures that can be applied: In the *centralized approach of cooperative positioning*, (Mayorga et al. 2007; Frattasi 2007) it is assumed that all information, that is, the measurements collected by the MTs, is provided to one central entity. That could be a location server in a wireless communications system. At this entity, the measurements are jointly processed and the position for each MT in the network is determined. Afterwards, this information can be exploited in the network or sent back to the MTs. As all measurements are processed jointly in this approach, it is the optimum procedure from a position estimation accuracy point of view. However, the main drawback is that all measurements have to be collected at a central entity in advance. So as to cope with scalability in dense large-scale networks or for MT-centric applications using restricted infrastructure, the *distributed cooperative positioning* approach can also be favored as an alternative to centralized methods, (Chan and So 2009; Wymeersch et al. 2009). Here, the MTs have only the information available that are obtained from their neighbors via P2P links and the measurements with the BSs. Hence, the position estimation complexity is distributed among the MTs compared to the centralized approach. An extensive overview of cooperative positioning techniques discussed under the framework of Bayesian inference can be found in (Wymeersch et al. 2009).

Generally, the communications overhead and extra-signalling is higher for cooperative approaches than for conventional, non-cooperative positioning. Furthermore, usually the overall overhead of distributed schemes is higher than for centralized schemes. Hence, signal-processing complexity and signalling overhead are two key problems for existing cooperative positioning approaches (see Mensing and Nielsen 2010; WHERE D2.4 2010). This problem can be significant especially for a wireless network accommodating a large number of MTs. Therefore, an efficient cooperative positioning scheme should achieve the best trade-off between communications overhead and position estimation performance.

7.2.2   Example for Centralized Cooperative Positioning

We consider N_{BS} BSs and N_{MT} MTs that are present in the scenario. The BSs are located at the known and fixed positions

$$x^{(BS)} = \left[x_1^{(BS),T} \; x_2^{(BS),T} \; \cdots \; x_{N_{BS}}^{(BS),T} \right]^T,$$

where

$$x_\mu^{(BS)} = \left[x_\mu^{(BS)} \; y_\mu^{(BS)} \right]^T, \quad \mu = 1, 2, \ldots, N_{BS},$$

describes the position of the BS μ. The positions of the N_{MT} MTs

$$x = \left[x_1^T \; x_2^T \; \cdots \; x_{N_{MT}}^T \right]^T$$

with

$$x_v = \left[x_v \; y_v \right]^T, \quad v = 1, 2, \ldots, N_{MT},$$

have to be estimated. Note that we restrict the derivation to a two-dimensional scenario, where an extension to three-dimensional approaches is straightforward.

The range between the MT v and the BS μ can be calculated as

$$r_{v,\mu}^{(MT-BS)}(x) = \sqrt{(x_\mu^{(BS)} - x_v)^2 + (y_\mu^{(BS)} - y_v)^2}$$

and the range between the MTs v and $v' \neq v$ is given as

$$r_{v,v'}^{(MT-MT)}(x) = \sqrt{(x_v - x_{v'})^2 + (y_v - y_{v'})^2},$$

where the dependence on the MT positions is explicitly denoted by x. An overview of the cooperative positioning principle with three BSs and three MTs is depicted in Figure 7.4. x_1 and x_2 have access to at least three other points of information, whereas x_3 is at least in the coverage range of x_1 (which might be able to estimate its position accurately).

The ranging error model for the MT-BS measurements can be written as

$$\hat{r}_{v,\mu}^{(MT-BS)} = r_{v,\mu}^{(MT-BS)}(x) + b_{v,\mu}^{(MT-BS)} + n_{v,\mu}^{(MT-BS)},$$

where the modeled bias $b_{v,\mu}^{(MT-BS)}$ and the residual noise $n_{\mu,v}^{(MT-BS)}$ depend on the LOS or NLOS status and the distance. Whereas the MT index $v = 1, \ldots, N_{MT}$ includes all MTs in the network, the BS index for each MT $\mu = 1, \ldots, N_{BS, Used,v}$ includes only the $N_{BS, Used,v} < N_{BS}$

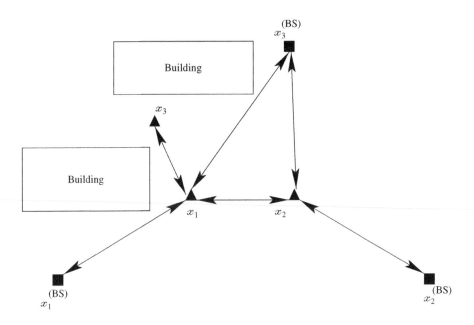

Figure 7.4   Cooperative positioning principle

BSs, which can be used for ranging from MT v. Equivalently, the ranging error model for the
MT-MT measurements is given as

$$\hat{r}_{v,v'}^{(\text{MT-MT})} = r_{v,v'}^{(\text{MT-MT})}(\mathbf{x}) + b_{v,v'}^{(\text{MT-MT})} + n_{v,v'}^{(\text{MT-MT})},$$

where $v' = 1, \ldots, N_{\text{MT, Used},v}$ includes the available other MTs of MT v for performing
ranging.

We include all available MT-BS and MT-MT measurements in the vector

$$\hat{\mathbf{r}} = \left[\hat{\mathbf{r}}^{(\text{MT-BS}),\text{T}} \ \hat{\mathbf{r}}^{(\text{MT-MT}),\text{T}}\right]^{\text{T}}$$

of dimension

$$N_{\text{Used}} = N_{\text{BS, Used}} + N_{\text{MT, Used}}$$

with

$$N_{\text{BS, Used}} = \sum_{v=1}^{N_{\text{MT}}} N_{\text{BS, Used},v}$$

and

$$N_{\text{MT, Used}} = \sum_{v=1}^{N_{\text{MT}}} N_{\text{MT, Used},v}.$$

With the equivalent definitions of the range vector $\mathbf{r}(\mathbf{x})$, the bias vector $\mathbf{b}$, and the noise vector
$\mathbf{n}$ with covariance matrix

$$\mathbf{\Sigma}_n = \begin{bmatrix} \mathbf{\Sigma}_n^{(\text{MT-BS})} & \mathbf{0} \\ \mathbf{0} & \mathbf{\Sigma}_n^{(\text{MT-MT})} \end{bmatrix},$$

we arrive at the compact measurement model

$$\hat{r} = r(x) + b + n.$$

For the static solution of the centralized cooperative positioning estimation problem, we follow the *weighted nonlinear least squares* approach (Gustafsson and Gunnarsson 2005; Kay 1993) according to

$$\hat{x} = \underset{x}{\arg\min}\,(\hat{r} - r(x))^T \Sigma_n^{-1}(\hat{r} - r(x)), \tag{7.1}$$

which was already discussed in Chapter 4. In the general case, there exists no closed-form solution to this nonlinear $2N_{MT}$-dimensional optimization problem, and hence, iterative approaches are necessary. A standard approach to deal with Equation (7.1) is based on the *Gauss–Newton algorithm* (Gustafsson and Gunnarsson 2005; Kay 1993). The Gauss–Newton algorithm linearizes the system model about some initial value $x^{(0)}$ yielding

$$r(x) \approx r(x^{(0)}) + \Phi(x)\,|_{x=x^{(0)}}(x - x^{(0)}),$$

with the elements of the $N_{Used} \times 2N_{MT}$ Jacobian matrix

$$\Phi(x) = \nabla_x^T \otimes r(x),$$

where

$$\nabla_x = \left[\frac{\partial}{\partial x_1}, \frac{\partial}{\partial y_1}, \cdots, \frac{\partial}{\partial x_{N_{MT}}}, \frac{\partial}{\partial y_{N_{MT}}}\right]^T.$$

Afterwards, the linear least squares procedure is applied resulting in the iterated solution

$$\begin{aligned} x^{(k+1)} &= x^{(k)} + (\Phi^T(x^{(k)})\Sigma_n^{-1}\Phi(x^{(k)}))^{-1} \\ &\quad \cdot \Phi^T(x^{(k)})\Sigma_n^{-1}(\hat{r} - r(x^{(k)})). \end{aligned}$$

The Gauss–Newton algorithm provides very fast convergence and accurate estimates for good initial values. For poor initial values and bad geometric conditions the algorithm results in a rank-deficient, and thus, non-invertible matrix for certain geometric constellations of MTs and BSs (see Chapter 4 for alternative static positioning algorithms).

For the considered approach, the initial position estimates of the individual MTs is defined by the mean value of the positions of the visible BSs, that is, corresponding to

$$x_v^{(0)} = \frac{1}{N_{BS,\,Used,v}} \sum_{\mu=1}^{N_{BS,\,Used,v}} x_\mu^{(BS)}.$$

Certainly, for moving MTs position tracking can be also applied straightforwardly. For instance, the EKF (see Section 5.2) can be extended for multiple MTs. The already discussed state-space and observation models are

$$s[k] = As[k-1] + u[k]$$

$$\hat{r}[k] = h(s[k]) + n[k], \tag{7.2}$$

where

$$s[k] = \begin{bmatrix} x_1^T & v_1^T & x_2^T & v_2^T & \cdots & x_{N_{MT}}^T & v_{N_{MT}}^T \end{bmatrix}^T$$

is now a $4N_{MT}$-dimensional state-space vector in each time-step $k \in \mathbb{N}$, including two-dimensional positions and velocities of each MT as parameters that have to be estimated. The vector $\hat{r}[k]$ includes the ranging measurements for each time step and changes over time depending on the availability of the measurements. The matrix

$$A = \left(1_4 + \left(\begin{bmatrix} T & 0 \\ 0 & T \end{bmatrix} \otimes \begin{bmatrix} 0 & 1 \\ 0 & 0 \end{bmatrix} \right) \right) \otimes 1_{N_{MT}}$$

includes *a priori* information about the MT movements with timing updates every T time-steps. The vector $u[k]$ is composed of state-space noise with the diagonal covariance matrix Q, and $n[k]$ is composed of the observation noise with the covariance matrix $\Sigma_n[k]$. The covariance matrix can change dynamically over time depending on number and type of available measurements. Finally, the function $h(\cdot)$ describes the nonlinear relation between the state-space vector and the measurements. The equations for the state-space and observation models in Equation (7.2) are then used to set-up the EKF, where the derivation from Section 5.2 can be directly applied.

7.2.3 Simulation Results

We assume an area of 100×100 m^2 including $N_{BS} = 20$ BSs and $N_{MT} = 24$ MTs. For modeling the ranging errors, we make use of the models presented in (WHERE D2.4 2010). They model bias and residual noise conditioned on distance, orientation, and LOS or NLOS status of the connection. It is assumed that the MT-MT connections are always LOS, whereas the MT-BS connections are NLOS in 50% of the cases. For emulating group mobility, a variation of a *random waypoint model* is exploited. In each group of MTs, one of the nodes acts as the reference or leading MT. For this MT a waypoint and speed (maximum of 2 m/s) is chosen as usual for the random waypoint model. For the remaining nodes in the group, the same speed is used and their waypoints are chosen so that they are randomly placed within 20 m of the reference MT's waypoint. An example of the resulting mobility tracks is shown in Figure 7.5 for the considered scenario. In this example, there are two groups with four nodes in each group. For the final simulations, six groups will be assumed.

Figure 7.6 shows the CDF for conventional, non-cooperative positioning and cooperative positioning for both static position estimation and position tracking by an EKF (see Mensing and Nielsen 2010; WHERE D2.4 2010). It can be observed that for the static approach more than 10% of the MTs cannot be localized, for example due to limited access to BSs or bad geometric conditions. This can be reduced by application of the EKF resulting in an error being smaller than 10 m in 90% of the cases. If cooperation between the MTs can be exploited, this can further be improved to around 3 m.

7.3 Multipath and Non-Line-of-Sight Mitigation

Positioning with classical GNSS is based on LOS propagation, that is, GNSS assumes a LOS path and NLOS propagation can usually not easily be detected due to too weak signals.

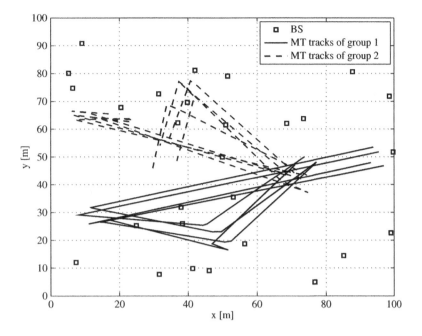

Figure 7.5   Group mobility principle

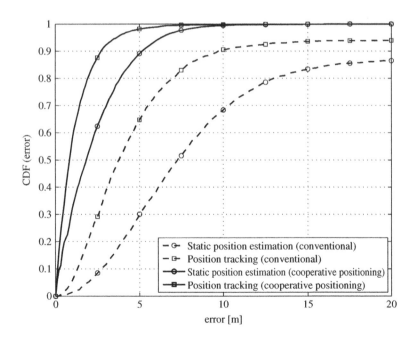

Figure 7.6   Cooperative positioning using static position estimation and position tracking with EKF

Contrary to that, positioning in wireless communications includes usually NLOS propagation. State-of-the art positioning as described before usually exploits time based information. Under NLOS propagation, a bias is introduced into the timing measurements performed between the MT and the BSs, which can have a remarkable effect on the overall positioning performance. For two-dimensional positioning, measurements with at least three BSs have to be performed, where for urban environments many of them might be under NLOS conditions. Propagation analysis in these environments for GSM systems have shown that the resulting NLOS error in the timing measurements can be in the order of several hundred meters. Also for current and future systems this NLOS error will be in similar regions. In addition to the NLOS error multipath propagation will also limit the positioning performance. Whereas the *NLOS error* is *always positive*, the *multipath error* can be *positive or negative*.

If it is not possible to mitigate these influences inside the timing estimation algorithms (e.g., by appropriate channel estimation algorithms), the resulting timing estimates have a positive or negative bias for multipath propagation and might have an additional positive bias under NLOS conditions. Two principles can be identified to cope with this problem. In the first, *a priori* knowledge of the bias statistics is required. This might also include the knowledge of the LOS or NLOS condition of the individual links between MT and BSs. It allows an appropriate weighting or setting of constraints to improve the overall location estimate (Cong and Zhuang 2005; Guvenc et al. 2010; Qi et al. 2006). A second idea is the exploitation of redundancy inside the available measurements. If more timing estimates are available than required, intermediate location estimates using only a subset of the available timing estimates can be computed. These intermediate estimates can then be combined to drop out outliers and to identify biased links (Chen 1999; Riba and Urruela 2004). Note that this concept is denoted as *receiver autonomous integrity monitoring (RAIM)* in the context of GNSS (Parkinson and Spilker Jr. 1996). A comprehensive overview about this topic can be found in (Guvenc and Chong 2009).

8

Systems

Investigating state-of-the-art and already standardized systems, for example, GSM, UMTS, and LTE, it becomes obvious that position information is not *a priori* part of the systems but has been seen more as an *add-on* to these communications systems. Hence, these systems have been designed for communications purposes only. However, encouraged by the US *Federal Communications Commission (FCC)* requirements and the demand for *GPS-free* location information in MTs, systems were extended to also allow stand-alone positioning.

In the following, we describe the standardized technologies in *second-generation (2G)*, *third-generation (3G)*, and *fourth-generation (4G)* systems. Telecommunications standards organizations worked on including location technologies into their standards, for instance for GSM, UMTS, LTE, cdmaOne, CDMA2000, *Wideband CDMA (W-CDMA)*, or WiMAX. Three main standard organizations involved in 2G systems were the *European Telecommunications Standards Institute (ETSI)*, *Telecommunications Industry Association (TIA)*, and the T1 Committee. T1 was sponsored by the *Alliance for Telecommunications Industry Solutions (ATIS)*. For 3G and 4G systems, the work has been handled by 3GPP, the *3rd Generation Partnership Project 2 (3GPP2)* and the *Institute of Electrical and Electronics Engineers (IEEE)*, respectively.

In this chapter, we will discuss several of the above and other systems and standards that are relevant for positioning in wireless communications. First, we present the 3GPP cellular communications systems GSM, UMTS, and LTE in Sections 8.1, 8.2, and 8.3. Afterwards we survey the IEEE systems WiMAX and WLAN (see Section 8.4). Then, we review the short range communications technologies Bluetooth, ZigBee, *ultra-wideband (UWB)*, *radio-frequency identification (RFID)*, and *near field communication (NFC)* in Section 8.5. Finally, we conclude the chapter with a short review on standards for positioning in wireless communications in Section 8.6.

8.1 GSM

We present in Sections 8.1.1 and 8.1.2 the system parameters and measurements that are relevant for positioning in GSM. Four positioning methods have been standardized in GSM (ETSI 2012a): *Timing advance (TA)* (Section 8.1.3), *enhanced observed time difference*

Positioning in Wireless Communications Systems, First Edition. Stephan Sand, Armin Dammann and Christian Mensing.
© 2014 John Wiley & Sons, Ltd. Published 2014 by John Wiley & Sons, Ltd.

EOTD) (Section 8.1.4), *uplink TOA (UTOA)* (Section 8.1.5), and *assisted GNSS (AGNSS)* (Section 8.1.6). Section 8.1.7 presents performance evaluations on different positioning methods.

8.1.1 System Parameters

In this section, we briefly present an overview on the system parameters of GSM that determine the achievable accuracies of the standardized positioning methods. Within Section 8.1, we will focus on the default GSM parameters summarized in Table 8.1 (ETSI 2012d), (ETSI 2012c), and (ETSI 2013a). Thus, we do not consider parameters for the *enhanced data rates for GSM evolution (EDGE)* or the *voice services over adaptive multi-user channels on one slot (VAMOS)*.

GSM can transmit in several frequency bands, among which GSM 850, 900, 1800, and 1900 are most widely used. GSM 850 and 900 operate in 25 MHz both in the downlink, BS to MT communication, and in the uplink, MT to BS communication, transmission. GSM 1800, also known as DCS 1800, and GSM 1900, also known as PCS 1900, operate in 75 and 60 MHz bands. The transmit power of the MT varies for GSM 850/900 between 5 and 33 dBm and for GSM 1800/1900 between 0 and 30 dBm in the uplink. The cellular network operators can use BSs with a maximum transmit power between between 34 and 58 dBm and between 34 and 46 dBm for GSM 850/900 and for GSM 1800/1900. The transmit power can be further decreased by 10 dB or optionally by 28 dB. The sensitivity of the MT and BS are -102 and -104 dBm.

The GSM symbol duration T_S is equal to $48/13\,\mu s \approx 3.69\,\mu s$. Thus, GSM transmits at a symbol rate of $1625/6\,kHz \approx 270.833\,kHz$. With 156.25 symbols per slot and eight slots per *time division multiple access (TDMA)* frame (see Figure 8.1), the TDMA frame is 4.6 ms long.

Table 8.1 System Parameters of GSM

Parameter	Value
Modulation	GMSK (bandwidth-time product $BT = 0.3$)
Carrier separation	200 kHz
Symbol duration	$\frac{48/13}{\mu s} \approx 3.69\,\mu s$
Symbol rate	$1625/6\,kHz \approx 270.833\,kHz$
Symbols/slot	156.25 symbols
Slot duration	576.92 µs
TDMA frame duration	4.6 ms
Maximum MT power	33 dBm (GSM 850/900),
	30 dBm (GSM 1800/1900)
Minimum MT power	5 dBm (GSM 850/900),
	0 dBm (GSM 1800/1900)
Maximum BS power	34–58 dBm (GSM 850/900),
	34–46 dBm (GSM 1800/1900)
Minimum BS power	10 dB or 28 dB below maximum BS power
MT sensitivity	-102 dBm
BS sensitivity	-104 dBm

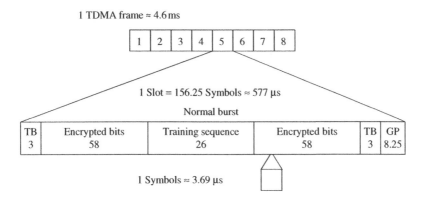

Figure 8.1   TDMA frame of GSM

Within each time slot, a burst can be transmitted. For instance, time slot 5 in Figure 8.1 contains a normal burst. The normal burst consists of three *tail bits (TBs)* at the burst start, followed by 58 encrypted data bits, a training sequence of 26 symbols, another 58 encrypted data bits, and at the burst end three more TBs and 8.25 symbols *guard period (GP)*. The midamble training sequence can be used for channel estimation and synchronization. Besides the normal burst, the GSM standard also defines a synchronization burst to enable time synchronization of MTs with a BS. The synchronization burst has an extended training sequence of 64 symbols and it contains information to compute the current TDMA frame number and the Cell-ID.

The default modulation scheme in GSM is *Gaussian minimum-shift keying (GMSK)*. GMSK is a special variant of partial response continuous phase modulation (Stüber 2001) and continuous-phase frequency-shift keying. It is a nonlinear modulation scheme with the following benefits:

- GMSK enables power efficient transmission with nonlinear amplifiers due to its constant-envelope waveform. This is in contrast to *quadrature phase-shift keying (QPSK)*, which requires more power efficient linear amplifiers as large envelope variations occur during phase transitions.
- GMSK reduces significantly the out-of-band power due to smooth phase transitions between adjacent symbols. For instance, Figure 8.2(a) displays the phase transitions within one slot both for GMSK and QPSK modulation. Clearly, QPSK has phase jumps of 90 or 180°, whereas the phase of GMSK changes smoothly. Note that the discontinuous phase changes for GMSK at ±180° are due to the mapping of the phase into the interval [−180°; 180°]. Due to the smooth phase transitions, the out-of-band power of GMSK in Figure 8.2(b) decreases substantially faster with increasing frequency compared to QPSK. Thus, neighboring frequency carriers from different BSs can be spaced closer to each other without increasing inter-cell interference, but increasing the overall system spectral efficiency. For instance, the carrier separation in GSM is only 200 kHz (ETSI 2012d).

While GMSK has the above benefits, it also has drawbacks. For instance, the Gaussian filter in GMSK creates inter-symbol interference. This requires equalization of the inter-symbol interference at the receiver, which increases the digital signal processing complexity.

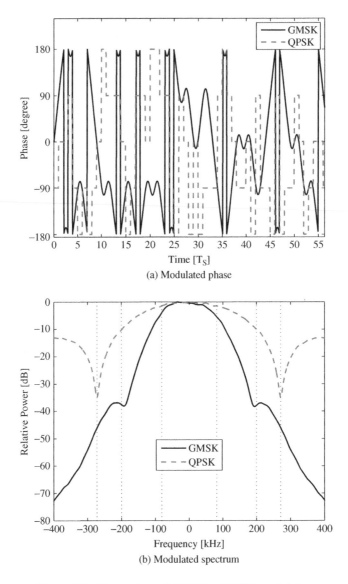

(a) Modulated phase

(b) Modulated spectrum

Figure 8.2   Comparison of GMSK and QPSK modulation

In Section 3.2, we introduced the relation between the CRLB for timing estimation and the signal spectrum (see 3.6). From the relation, it is clear that larger signal bandwidths and signal energy at the band edges will increase the timing estimation accuracy. Hence, the timing estimation accuracy of narrowband GMSK signals as in Figure 8.2 will be poor. We provide more detail on this in Section 8.1.7.

8.1.2   Measurements

The GSM standard specifies the following four distance related measurements:

- *Root-mean-square received signal (RMS-RX) level*: This value is mapped to the RXLEV parameter in GSM, which is used in the handover process of an MT between different BSs (ETSI 2012e). The RMS-RX level can be measured with the training sequences in the TDMA burst (see Figure 8.1) in the current up- and downlink channels and in the *broadcast control channel (BCCH)* carrier. The BCCH is continuously transmitted by all BSs in all time slots with constant transmit power. The RMS-RX level at the receiver input shall be measured by the MT in the range of −110 to −38 dBm and by the BS in the range of −110 to −48 dBm with the accuracies specified in Table 8.2. For the RMS-RX level, no positioning method has been standardized in GSM. As the RMS-RX level is used in the handover process, the MT measures the RMS-RX level of the serving BS and of up to six strongest neighbor BSs. Thus, the RMS-RX level measurements are available to the MT and BS allowing RSS based localization of the MT (see Sections 2.3.2, 2.3.3, and 2.3.4).
- *Timing advance*: This measurement specifies the round-trip delay time and is used to ensure synchronous uplink transmission of MTs within an TDMA frame (see Section 8.1.3).
- *Observed time difference (OTD)* and *real time difference (RTD)*: These measurements are used for pseudo-synchronization, synchronized handovers and pseudo-synchronized handovers as well as in the *enhanced observed time difference positioning method* (see Section 8.1.4).
- *TOA*: These measurements are used in the uplink TOA positioning method for GSM (see Section 8.1.5).

In the subsequent GSM sections, we present the standardized positioning methods.

8.1.3   Timing Advance – TA

For GSM, the simplest stand-alone location technology is the timing advance method. To understand this method, we first present the GSM multiple access scheme. GSM uses *TDMA* to share a single frequency channel between several users both in the downlink and in the uplink. It assigns sequential timeslots to the individual MTs sharing a single frequency (see Figures 8.1 and 8.3). A TDMA frame consists of eight time slots. Thus, a maximum of eight MTs can receive and transmit data within one TDMA frame.

Table 8.2   RMS-RX level and accuracies

RMS-RX level [dBm]	Accuracy [dBm]
−110 to −70	±4 at MT and BS
−70 to −48	±6 at MT and BS
−48 to −38	±9 only at MT

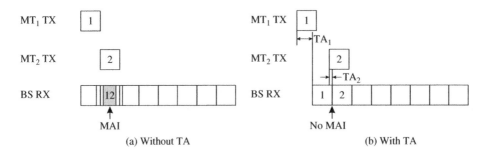

Figure 8.3 Timing advance in GSM

As the different MTs are at different locations, the signal propagation times between a BS and each MT varies with the distance. Therefore, the timing advance method has been introduced in GSM to ensure synchronous transmission in the uplink and to avoid *multiple access interference (MAI)* as shown in Figure 8.3.

The timing advance can be determined and tracked by the BS by measuring the delay of each received burst relative to the clock of the BS. To obtain the delay, the BS can, for example, correlate the received burst with the known training sequence contained in the burst. The timing advance is specified in multiples of 1 symbol duration, that is, 3.69 µs. As GSM requires the uplink transmission to be synchronous with the downlink transmission, the timing advance specifies the round-trip delay time. Thus, it offers a spatial accuracy of 553 m. In GSM, the timing advance can vary between 0 and 63. Hence, the timing advance covers a maximum distance of approximately 35 km, which corresponds to the maximum GSM cell radius. A position estimate based on the timing advance in combination with the Cell-ID or cell sector identity consequently results in a ring or ring segment with a distance accuracy of 553 m.

8.1.4 Enhanced Observed Time Difference – EOTD

For GSM, the main stand-alone location technology is based on TDOA and called *EOTD*. EOTD has been finalized by the GSM standard committees in *location services (LCS)* Release 98 and Release 99 (3GPP 2001). EOTD is a TDOA positioning method based on the OTD feature that is a part of GSM for synchronized or pseudo-synchronized handovers. There, the MT measures the relative time of arrival of the signal bursts from several BSs. The position of the MT is then determined by hyperbolic positioning techniques. Generally, there are three timing quantities involved in this method (see Figure 8.4):

- OTD is the time interval observed by an MT between the receptions of signal bursts from two different BSs.
- RTD is the relative synchronization in the network between two related BSs.
- The *Geometric Time Difference (GTD)* is the time interval measured at the MT between bursts from two BSs due to the overall geometry.

Hence, we have the relation between the involved time differences given as

$$GTD = OTD - RTD.$$

This relation is depicted in Figure 8.4(a).

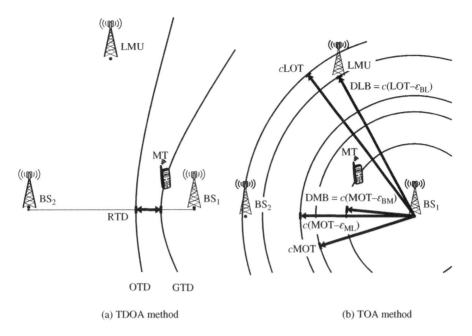

(a) TDOA method   (b) TOA method

Figure 8.4   EOTD in GSM

The OTD can be measured at the MT by detecting the arrival time of the training sequences contained in the bursts transmitted from several BSs. For synchronized networks the RTD is equal to zero, and hence the GTD equals OTD. For non-synchronized networks the RTD could be measured by so-called *location measurement units (LMUs)* that are part of the infrastructure. The LMUs thus can determine the asynchronism of the network and provide this to the MT.

The measurement accuracy of EOTD is in the best case $1/256$ symbol duration, that is, 14.4 ns, for the OTDs and RTDs (ETSI 2012b). Thus, the measurement resolution is approximately 4.32 m with a minimum uncertainty between 0–9 m. Furthermore, the MT can measure a timing advance correction with an resolution of $1/64$ symbol duration, that is, 17.3 m. Note that measuring the timing advance and one OTD with these accuracies is similar to measuring two TOAs from the reference BS and a neighboring BS resulting in two location estimates. To determine the correct MT location, another independent measurement, for example, a further OTD, is required.

Another method in the framework of EOTD measures the time of arrival of the signals from a BS to the MT and to the network node *LMU*. The following quantities are used:

- MOT: Observed time from a BS to the MT as time measured against the internal clock of the MT.
- LOT: Observed time from a BS to the LMU as time measured against the internal clock of the LMU.
- ϵ_{BM}: Time offset as bias between the two internal clocks of the MT and BS.
- ϵ_{BL}: Time offset as bias between the two internal clocks of the BS and LMU.

- ϵ_{ML}: Time offset as bias between the two internal clocks of the MT and LMU, which is given by

$$\epsilon_{ML} = \epsilon_{BM} - \epsilon_{BL}. \qquad (8.1)$$

- DMB: Distance from MT to BS.
- DLB: Distance from LMU to BS.

The relation of the different quantities is shown in Figure 8.4(b). Hence, we have the relation between the involved measurements given as

$$DMB - DLB = c(MOT - \epsilon_{BM} - LOT + \epsilon_{BL})$$
$$= c(MOT - LOT - \epsilon_{ML}),$$

where we used (8.1) to obtain the final result.

For each BS, one of these relations can be found. Since there are three unknowns included, two-dimensional MT position and clock offset ϵ_{ML}, at least three BSs are required to find a unique solution: see Section 4.2.

EOTD in general requires a minimum of three spatially distinct BSs where all of these BSs must be detectable by the MT. More than three measurements generally produce better location accuracies. An implementation of the EOTD method requires an LMU to BS ratio between 1:3 or 1:5 (3GPP 2001).

8.1.5   Uplink Time of Arrival–UTOA

Also a TOA procedure has been specified and standardized for GSM. It is the so-called network based UTOA method.

This method is based on measuring the TOA of a known signal burst from the MT at three or more *LMUs* in the infrastructure. The known signal is the training sequence contained either in a normal burst or in an access burst generated by the MT to perform an asynchronous handover. Compared to the normal burst (see Figure 8.1), the access burst contains eight tail bits, a 41 symbol long training sequence, 36 data bits, three tail bits and a 68.25 bit long guard period. The longer training sequence and guard period of the access burst have the advantage that LMUs located outside the serving cell can measure the TOA of the MT more accurately. The LMUs can measure and report received TOAs from the MT with an accuracy of 4 ns.

After the signal measurements at the network nodes, the TDOA approach is used to determine the position of the MT. Hence, it calculates the time difference of at least two pairs of TOA signals and derives the MT position by hyperbolic techniques according to TDOA. Therefore, UTOA is a hybrid method combining TOA and TDOA. The position calculation or solution of the navigation equation can be done in the same way as shown for EOTD. The only difference is that the computation is performed in the infrastructure in network based UTOA, whereas in EOTD it is done in the MT.

It was shown that UTOA is more effective at reducing noise and interference through correlation and burst averaging than EOTD (3GPP 2001).

8.1.6   Assisted GNSS–AGNSS

GNSS are satellite systems that have been or are going to be deployed for positioning and navigation purposes (see Section 1.2). Systems belonging to this category are operational today or will be in the near future, for example GPS and its modernization, GLONASS, Galileo, WAAS, EGNOS, MSAS, and GAGAN. AGNSS approaches were specified to improve the performance of GNSSs:

- Faster time to first fix.
- Better reception of weak signals in urban canyons or indoor especially for cellular-sized antennas.
- Reduced power consumption by reducing signal acquisition time.

The main idea of AGNSS is to set up a GNSS reference network or equivalently a wide-area DGNSS network, where the receivers have clear sky-views. This reference network is additionally connected with the cellular infrastructure, continuously monitors the real-time constellation status, and provides data such as approximate MT position (or BS location), satellite visibility, ephemeris, and clock correction, Doppler, and the code phases for each satellite. On demand the assistance data is then transmitted to the MT or network nodes for fast start-up and increased sensitivity. When the available satellite signals are detected, the pseudo-range measurements can be delivered to the network for position calculation or used internally in the MT to compute its position. Additional assisted data, such as real-time integrity, DGNSS corrections, satellite almanac, ionospheric delay, and *coordinated universal time (UTC)* offset can be transmitted.

There are two fundamental modes that are supported:

- The *MT-assisted* solution shifts the majority of the GNSS receiver functions to the network processor. This method requires at least an antenna, RF section, and base-band processor in the MT for making measurements by generating replica codes and correlating them with the received GNSS signals. By AGNSS an assistance message is sent to the MT, consisting of time, visible satellite list, satellite signal Doppler, and code phase, as well as their search windows or, alternatively, approximate handset position and ephemeris. The assistance data of Doppler and code phase is usually valid for a few minutes, while ephemeris data is valid for around 2–4 hours. From the MT the pseudo-range data is returned to the network. Then, the location server estimates the position of the MT. Additionally, differential correction in terms of DGNSS can be applied to the pseudo-range data or final result at the network side to improve the position accuracy.
- For *MT based* solutions, a complete GNSS receiver is integrated in the handset. In the start-up phase, satellite orbital elements, that is, the ephemerides, must be provided to the MT. For instance, in the case of GPS, this data is valid for 2–4 hours and can be extended to cover the entire visible period of the GPS satellite, that is, up to 12 hours. For better positional accuracy or longer ephemeris life, differential correction data from DGNSS could be transmitted to the MT. The final position of the MT is generated at the MT. Then, the estimated MT position can be sent to the network if required.

8.1.7   Cramér–Rao Lower Bounds

In this section, we present accuracy estimations based on the CRLB for GSM and different measurement types.

8.1.7.1   Cellular Setup

The cellular network is setup in accordance to Figure 8.5. Each cell of the $N_{BS} = 7$ BSs is divided in three hexagonal sectors by three sectorized antennas. The sectorized antennas (see Figure 8.6) cover approximately an angle of 120°, have a 3 dB beamwidth of 70° and a maximum antenna gain of 14 dBi. Each BS transmits per sector with a maximum power of 43 dBm. The inter-site distance between BSs is 750 m. The MT position is randomly located within the gray shaded target sector S1 of the central, serving BS in Figure 8.5. To obtain the CRLBs in the subsequent figures, $5 \cdot 10^5$ random MT positions were drawn within sector S1. In this section, the GSM system transmits signals at a carrier frequency of $f_c = 2$ GHz for better comparability with other subsequently presented systems. The minimum RX sensitivity of the MT was set to -102 dBm (see Table 8.1).

8.1.7.2   Channel Model

For our analyses and simulations, we need a channel model that provides multiple-link capabilities. Such capabilities are provided by channel models, developed during the EU-Project WINNER (IST-2003-507581 WINNER 2007). The WINNER models are based on clusters of scatterers for multipath modeling. Besides small scale fading, that is, frequency selective fading, these models also cover large scale effects like delay spread/distribution, angle of departure spread/distribution, angle of arrival spread/distribution, shadow fading, and Rice factor.

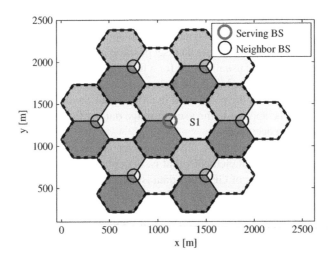

Figure 8.5   Cellular network layout: One serving cell with six neighbor cells and three hexagonal sectors per cell

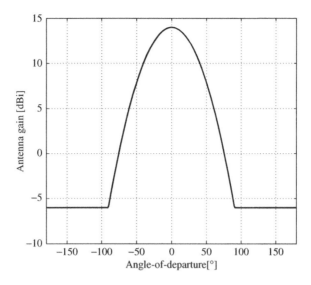

Figure 8.6 BS transmit antenna pattern

The parameters of these effects strongly depend on the propagation environment and have been modeled for a variety of environmental scenarios. Parameters, supported by the WINNER models are scaling of delay distribution, cross polarization, number of clusters, cluster angle spread of arrival/departure, per cluster shadowing, autocorrelation of large scale parameters, cross-correlation of large scale parameters, and number of rays per cluster.

For our simulations, we use the WINNER II typical urban macro cell channel model C2 (IST-2003-507581 WINNER 2007). This model describes a typical urban macro cell for mobile velocities of 0–120 km/h and carrier frequencies in the range of 2–6 GHz. In this scenario, the mobile station is located outdoors at street level. It provides a mixture of LOS and NLOS propagation conditions, where the LOS probability depends on the distance between the MT and BS. The LOS or NLOS propagation condition state of a link itself influences further parts of the model like the path loss model for instance. In the following section, the C2 channel model generates realistic predictions for PL, SF, and multipath for each MT position (IST-2003-507581 WINNER 2007).

8.1.7.3 Results

Figure 8.7 plots the *CDF* of the positioning CRLB $P[CRLB(x) \leq \sigma_x]$ vs. σ_x [m], that is, the probability that the positioning CRLB(x) is smaller than the standard deviation σ_x on the abscissa. As the derivation of CRLBs assume unbiased estimators, σ_x is equivalent to the RMSE (see Section 4.5.1). Note that the positioning CRLB analysis should be understood as a best case scenario in terms of positioning accuracy because we assume perfect synchronization of all BSs, a relatively good BSs geometry for positioning, interference-free reception of the signals from different BSs, and perfectly known multipath channel parameters. The known multipath channel parameters include the amplitudes and relative delays as well as any additional delays due to NLOS propagation on the first received path. In practice, these

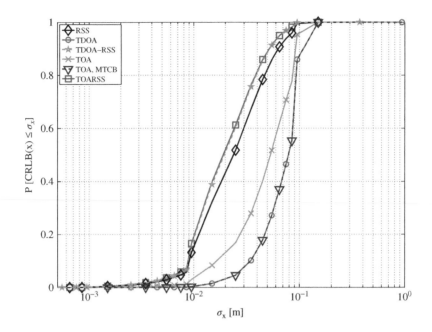

Figure 8.7   CDF of two-dimensional positioning CRLB with different measurements for GSM

assumptions cannot be met perfectly and hence, practical positioning errors may be two to four orders of magnitudes larger than these CRLBs. Nevertheless, the presented analysis allows to compare different measurement types and communications systems.

For these assumptions, the standard deviation of the positioning error is almost surely below 15 cm for all measurement types. As GSM transmits narrowband signals (see Figure 8.2b), the positioning errors of *TDOA* and *TOA* measurements with or without *MT clock bias (MTCB)* are larger than the ones of *RSS*, TDOA-RSS, and TOARSS measurements. The hybrid TOARSS measurements yield the best accuracy followed by the hybrid TDOA-RSS measurements with nearly the same accuracy, although the MTCB can be non-zero for the later one. Note that TOARSS or TDOA-RSS are measurements that jointly measure the TOA and RSS or the TDOA and RSS (Sand et al. 2012). As expected, the CRLBs for TDOA and TOA measurements with MTCB coincide (Urruela et al. 2006). For a CEP of 67% or 95% (see Section 4.5.3), the positioning error for TOARSS or TDOA-RSS is below 2.9 and 6.6 cm and for TDOA below 8.9 and 13 cm. Thus, the TOARSS or TDOA-RSS measurements achieve an improved accuracy of a factor of two to three compared to the TDOA only measurements for GSM (see Table 8.3).

8.2   UMTS

In the *UMTS terrestrial radio access network (UTRAN)*, the MT is called *user equipment (UE)* and the BS is called *node B*. Generally, there are two operational modes for

Table 8.3 CEP$_x$ of positioning CRLB for GSM and different measurements (units in cm)

CEP$_x$	RSS	TDOA	TDOA-RSS	TOA	TOARSS
$x = 67\%$	3.5	8.9	2.9	7.1	2.9
$x = 95\%$	8	13	6.6	9.5	6.5

UTRAN: *frequency-division duplex (FDD)* and *time-division duplex (TDD)*. The original standards specifications were developed based on the FDD mode.

Four location techniques have been specified for UTRAN (ETSI 2012j): *Cell-ID* based, *observed TDOA (OTDOA), uplink TDOA (UTDOA)*, and *AGNSS* methods. According to (TSG-RAN 2001), all these methods are optional in the MT except for the MT-assisted OTDOA method. Next, we explain in detail the Cell-ID and *enhanced Cell-ID (E-CID)* based positioning methods, and the OTDOA positioning method. As the AGNSS and UTDOA methods do not differ significantly between GSM and UMTS, we omit details on these two methods and refer the reader to Sections 8.1.6 and 8.1.5.

8.2.1 System Parameters

In this section, we briefly present an overview on the system parameters of UMTS that determine the achievable accuracies of the standardized positioning methods. Within Section 8.2, we will focus on the default UMTS FDD parameters summarized in Table 8.4 (ETSI 2013i). Where necessary, we provide information on deviating parameters for UMTS TDD.

UMTS FDD can transmit in 17 paired frequency bands (ETSI 2013l). Among these, the UMTS FDD bands I, II, IV, V, and VIII listed in Table 8.5 are most widely deployed (Wikipedia 2013b). In contrast to GSM, UMTS employs linear modulation schemes, that is, *M-ary PSK* or *quadrature amplitude modulation (QAM)*. The transmit symbols are filtered with a *root-raised cosine pulse* with a roll-off factor of 0.22 at the chip rate of 3.84 Mcps. Figure 8.8 depicts the resulting average modulated spectrum for UMTS FDD. The vertical dashed lines at ± 2.5 and ± 1.92 MHz mark the halved carrier separation frequency and the 3 dB cut-off frequency. At ± 2.5 MHz, the UMTS FDD spectrum is dampened by more than 35 dB. Concerning the modulated spectrum of UMTS TDD, the only change compared to UMTS FDD is the chip rate, which can be either 1.28, 3.84, or 7.68 Mcps. Thus, for UMTS TDD the carrier separation and main signal bandwidths are 1.6, 5, and 10 MHz.

UMTS employs spreading to enable multiuser communication through CDMA. The spreading code length varies between 4 and 512 chips in the downlink. With a chip rate of 3.84 Mcps, the chip duration is 260.42 ns. An UMTS FDD frame consist of either 256 or 38 400 chips having a duration of 66.7 µs or 10 ms. According to the UMTS specifications, the maximum MT transmit power is ranging from 21 up to 33 dBm depending on the UMTS frequency band. The minimum MT transmit power is below 50 dBm. The maximum receiver sensitivity of the MT is defined between -117 and -114 dBm depending on the frequency band. For the transmit power and the maximum receiver sensitivity of BSs, the UMTS standard defines three classes that are wide are BSs, medium range BSs, and local area or home BSs.

Table 8.4   System parameters of UMTS FDD

Parameter	Value
Modulation	QPSK, 8-PSK (1.28 Mcps option TDD), 16-QAM (HSDPA, HSPA+), 64-QAM (HSPA+)
Pulse shape	Root-raised cosine
Roll-off factor	0.22
Carrier separation	5 MHz
Spreading code length	4–256 chips, downlink also 512 chips
Chip duration	260.42 ns
Chip rate	3.84 Mcps
Chips/frame	38 400 or 256
frame duration	10 ms or 66.7 μs
Maximum MT power	21–33 dBm (UMTS 2100)
	21–24 dBm (all other bands)
Minimum MT power	< 50 dBm
Maximum BS power	no upper limit for wide area BS
	38 dBm medium range BS
	24/20 dBm local area/home BS
Minimum BS power	at least 28 dB below
	maximum BS power
MT sensitivity	−117− − 114 dBm
BS sensitivity	−121 dBm wide area BS
	−111 dBm medium range BS
	−107 dBm local area/home BS

Table 8.5   UMTS FDD widely deployed frequency bands

Operating band	Frequency band [MHz]	Uplink frequencies [MHz]	Downlink frequencies [MHz]
I	2100	1920–1980	2110–2170
II	1900	1850–1910	1930–1990
IV	1700	1710–1755	2110–2155
V	850	824–849	869–894
VIII	900	880–915	925–960

8.2.2   Measurements

The UMTS standard specifies the following four distance dependent measurements:

- *Received signal code power (RSCP)*: These measurements are used for handover proce-dures, downlink or uplink open loop power control, as well as for calculations of the received energy per chip per power density in the band, that is, the *signal-to-interference-and-noise ratio (SINR)*, and pathloss (ETSI 2013j; ETSI 2013k). For FDD, the *common pilot channel (CPICH)* RSCP measures the received power of one code on the primary CPICH,

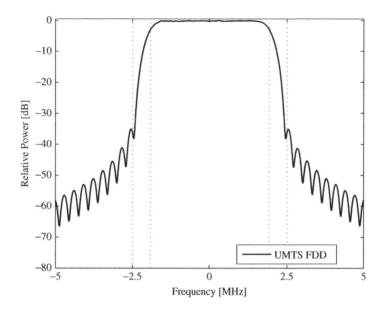

Figure 8.8   UMTS-FDD modulated spectrum

whereas for TDD, the *primary common control physical channel (P-CCPCH)* RSCP measures the received power of one code measured on the P-CCPCH. The UTRAN RSCP measures only for TDD at the BS the received power on one code in the *dedicated physical channel (DPCH), physical random access channel (PRACH), physical uplink shared channel (PUSCH)*, and so on (ETSI 2013k). The CPICH, P-CCPCH and UTRAN RSCP measurements are mapped to the CPICH_RSCP_LEV, the P-CCPCH_RSCP_LEV, and the RSCP_LEV parameters in UMTS. The RSCPs at the receiver input shall be measured in the range of −120 to −25 dBm for CPICH and P-CCPCH and in the range of −120 to −57 dBm for UTRAN with the accuracies specified in Table 8.6. For the RSCP measurements, the E-CID positioning method has been standardized in UMTS (see Section 8.2.3).

- *OTD*: The SFN-SFN OTD Type 2 is used for location service purposes to identify time difference between two cells (ETSI 2013j) with respect to the *system frame number (SFN)*. The SFN-SFN OTD Type 2 is measured by the UE with an accuracy of ±0.5 or ±1 chip for RSCPs larger than −94 dBm. The SFN-SFN OTD Type 2 is mapped to the T2_SFN-SFN_TIME parameter in the range of −1280–1280 chips with a step size of 0.0625 chips. This is equivalent to a time resolution of 16.3 ns or a spatial resolution of 4.9 m. The OTD positioning method has been standardized in UMTS (see Section 8.2.3). Additionally, *LMUs* can measure SFN-SFN OTD in the UTRAN to estimate RTDs.

- *Round-trip time (RTT)* and *PRACH propagation delay*: For FDD, (ETSI 2012h) standardizes the RTT and PRACH propagation delay. The RTT measurement of a BS together with the *UE Rx-Tx time difference* Type 2 measurement, which corresponds to the processing time of the UE, provides the round-trip delay estimate. The PRACH propagation delay is defined as the one-way propagation delay as measured during PRACH access. The RTT measurements have an accuracy of ±0.5 chip and the PRACH propagation delay measurements one of ±2 chips. The RTT measurements are reported with a step size of 0.0625

Table 8.6 RSCP level and accuracies under normal conditions

Channel	RSCP [dBm]	Accuracy [dBm]
CPICH & P-CCPCH	−94 to −70	±6
	−70 to −50	±8
UTRAN DPCH, PRACH, PUSCH, etc.	−105 to −60	±6 only at BS

chips corresponding to a spatial resolution of 4.9 m. In contrast to the RTT measurement (see Sections 8.2.3 and 8.2.4), the PRACH propagation delay measurement is currently not included in the standardized positioning methods for UMTS. For TDD, the *TA* or *RX timing deviation* measurements may be used instead of the UE Rx-Tx time difference Type 2 to calculate the round-trip delay.

- *AOA*: For the 1.28 Mcps TDD mode, (ETSI 2012i) defines the angle of arrival measurement as the estimated angle of a user with respect to a reference direction. The reference direction for the AOA measurement shall be the north, positive in a counter-clockwise direction. The AOA is determined at the BS antenna for an uplink channel corresponding to the MT. Depending on the UTRAN accuracy class (ETSI 2013k), the AOA measurement accuracy varies between ±1 to ±180° and is reported in 0.5 degree steps. The AOA measurement has been included in the standardized E-CID and OTDOA positioning methods for UMTS (see Sections 8.2.3 and 8.2.4).

In the following two sections, we explain in detail the Cell-ID and E-CID based positioning methods and the OTDOA positioning method. As the AGNSS and UTDOA methods do not differ significantly between GSM and UMTS, we omit details on these two methods and refer the reader to Sections 8.1.6 and 8.1.5.

8.2.3   Cell-ID and Enhanced Cell-ID Based

Using the Cell-ID approach, the MT position can be roughly determined in the network as specified for 3GPP. In other words, the position of a MT is estimated based on the coverage information of the serving BSs. This knowledge could be obtained by paging, locating area update, and cell update, UTRAN registration area update, or routing area update. Even though this method is optional for the network, this approach should be implemented as the default location method. Then, whenever OTDOA or AGNSS fail to locate the MT, one can always use this method to provide approximate information on MT position to the network.

Depending on the operational status of the MT, additional operations may be needed in order for the serving BS to determine the Cell-ID. When the location service request is received from the core network, the serving BS checks the state of the target MT. If the MT is in a state where the Cell-ID is not available, the MT is paged so that the serving BS can establish the cell with which the target MT is associated. In states where the Cell-ID is available, the target Cell-ID is chosen as the basis for the MT positioning. In case of soft handover, the MT may have several signal branches connected to different cells while reporting different Cell-IDs. The serving BS needs to combine the information about all cells associated with the MT to determine a proper

Cell-ID. Finally, the serving BS should also map the Cell-ID to geographical coordinates or a corresponding service area identity before sending it from the UTRAN to the core network. This can easily match the service coverage information available in the core network.

Hence, the Cell-ID based approach can determine the position of the MT regardless of the MT operational mode, that is, connected or idle. However, this method provides a position error as large as the cell area if no additional measurements are used. For instance, a small pico-cell could be 150 m in radius, while a large cell could be more than 30 km in radius.

In order to improve the accuracy of the location service, the serving BS may also request additional measurements from the BSs or *LMUs* (ETSI 2012j). These measurements are originally specified for soft handover and have been now defined as *E-CID* methods. For FDD mode, *RTT* can be used as a radius of a cell to further confine the cell coverage. RTT is the time difference between the transmission of the beginning of a *DPCH* frame to a MT and the reception of the beginning of the corresponding uplink from the MT. For better accuracy in FDD mode, a mandatory MT receiver-transmitter time difference Type 1 or an optional Type 2, if available, can be combined with RTT to determine the distance from a BS to a MT. For Type 1, the MT receiver-transmitter time difference is the difference between the MT uplink *dedicated physical control channel (DPCCH)* or *dedicated physical data channel (DPDCH)* frame transmission and the first detected path in time of the downlink DPCH frame from the measured radio link. The main differences in Type 1 and Type 2 are the measurement resolution and reference receiver path or first detected path. For Type 2, the resolution is better, that is, ± 1 chip accuracy instead of ± 1.5 chip accuracy. Further, the reference path must be the first detected path among all paths detected by the MT. In contrast, for Type 1, the reference path is the one used in the demodulation process.

For TDD mode, *received timing deviation* can be used. Received timing deviation is the time difference between the reception in the BS of the first detected uplink path and the beginning of the respective slot according to the internal timing of this BS. The measurements are reported to higher layers, where timing advance values are calculated and signaled to the MT, which is similar to the timing advance in GSM. Since adaptive or smart antennas have been specified as a feature for the BS, *AOA* can be used to further improve the Cell-ID performances in the 1.28 Mcps TDD mode (ETSI 2012j).

In addition to *E-CID with RTT*, (ETSI 2012j) also specifies Cell-ID with *RSS* measurements, that is, Cell-ID with *pathloss* and related measurements, or *RF pattern matching (RFPM)*. MT measurements, which could be used, are the *CPICH RSCP*, the CPICH received energy per chip divided by the power density in the band (E_c/N_0), and the pathloss measured in dB. RFPM uses, for instance, both RSS based measurements such as CPICH-RSCP and timing based measurements such as RTT in addition to the Cell-ID. Whereas Cell-ID with pathloss and related measurements relies on an analytical relationship between measurements and the distance dependent radio propagation model, Cell-ID with RFPM does not assume a simple analytical relationship between these (see Section 2.3.3). Instead, it searches in a database for the best match that provides the most likely position.

8.2.4   Observed Time Difference of Arrival–OTDOA

For UMTS, a TDOA positioning method is also specified by 3GPP. In OTDOA the position of the MT is determined by trilateration (see Section 4.2). Two methods are specified for OTDOA: MT-assisted OTDOA and MT based OTDOA.

Since these measurements are based on the signals from BSs, the locations of these BSs are necessary for the network or MT to calculate the MT positions. If the transmitters in UTRAN are non-synchronized, the *RTD* must be provided, similar to 2G systems by SFNs. It is named *SFN-SFN OTD*. One way to obtain these measurements is again to deploy *LMUs*, which perform timing measurements of all the local transmitters in fixed locations of the network. These measurements can be converted to RTDs and transmitted to the MT or network's location entity for positional calculations. In addition, the MT also measures the SFN-SFN OTD, which identifies the time difference between two cells as TDOA. Two types are defined. Type 1 is used for soft handover and Type 2 is used for positioning. The main difference of these two types is that Type 2 is applicable for both idle and connected modes, while Type 1 supports intrafrequency measurements and cannot do interfrequency measurements for the connected mode. Since BSs in TDD mode are generally synchronized, the RTDs are typically constant. Similarly, in FDD if the relevant cells are synchronized, measurements of RTDs are not necessary. For the FDD mode, RTT and receiver-transmitter time differences can be obtained to improve the performance of the OTDOA measurements. For the TDD mode, receiver timing deviation can be obtained to improve the performance. Since adaptive or smart antennas have been specified as a feature, AOA can be used to further improve OTDOA (ETSI 2012j).

The OTDOA location method in UTRAN has its problems, such as hearability, non-synchronized BSs for FDD mode, and bad relative geometric locations of the BSs. For the hearability problem, this can occur when the MT is very close to its serving BS, which could block the reception of other BS signals in the same frequency. This can be a serious problem since the MT must be able to hear at least three BSs in order to perform the location process. For the non-synchronized BS problem, it causes significant uncertainty to the TDOA measurements. For the geometric location of the BS problem, the locations of the contributing BSs could affect the availability and quality of the position fixes. For instance, on a long straight highway, OTDOA may fail to produce required solutions because the BSs may simply lie along a line.

In order to improve the hearability of neighboring BSs, one specified option is the *idle period downlink (IPDL)*. In this method, each BS interrupts its transmission for short periods of time, that is, idle periods. During an idle period of a BS, MTs within the cell can measure other BSs, so the hearability problem is reduced. By using signaling between the BSs and MTs, MTs are made aware of the occurrences of IPDLs, so they can arrange the time difference measurements accordingly. Since the IPDL method is based on downlink, the location service can be provided efficiently to a large number of MTs simultaneously.

For MT-assisted OTDOA, essential information elements or assistance data from UTRAN to MT are reference and neighbor cell information. For MT based OTDOA, they are reference and neighbor cell information as well as BS positions of these cells. MT-assisted OTDOA is mandatory for the MT and optional for the UTRAN. The MT based OTDOA is optional for both the MT and UTRAN. Note that for MT based OTDOA, the length of the downlink information elements is longer than that for the MT-assisted OTDOA. In contrast, the length of the uplink information elements is shorter for MT based OTDOA as it reports only the MT position whereas MT-assisted OTDOA reports measured TDOA results.

8.2.5 Comparison of UMTS and GSM

Comparing GSM and UMTS, similar features are included in the 3G standard as in the 2G standard. To reduce infrastructure investment, a shared location server could be implemented to support MTs operating in both 2G and 3G networks.

However, the cellular time base is in general different for GSM and UMTS. In GSM, cellular network time can be expressed in terms of broadcast control channel carrier, BS identity code, frame number, time slot number, and bit number. For UMTS, it can be expressed in terms of UTRAN-GPS timing of cell frames, primary common pilot channel info, and SFN. The uncertainty between cellular and GPS time is included in the GSM field *GPS Reference Time Uncertainty* and in the UMTS field *SFN-Time-Of-Week Uncertainty*. In UMTS, additionally the drift rate between network time and cellular time can be provided.

Since GSM and UMTS MTs do not have precise time information available internally, methods are included in the standards protocol to deliver precise time to the GSM/UMTS handset (ETSI 2012j). To accomplish precise time transfer in asynchronous GSM or UMTS networks, LMUs can be used. For AGNSS, the LMU measures the relationship between cellular frames from the serving BS with respect to GNSS time and sends this information periodically to the network server. The network server collects the time stamp information, and maintains a database of the relationship between cell timing and GNSS time for every BS. This information is then sent to the MT. Nevertheless, installing LMUs in the network is rather expensive for a network operator. Therefore, the GSM or UMTS standard allows the MT to perform the LMU function (ETSI 2012j). In that sense, after a position request, the MT reports the relation between cellular time and GNSS time to the network. The network uses this information to build up the database of cell timing and uses this information for assistance of other MTs. Hence, also in GSM or UMTS it is possible to substantially reduce the contribution of time error.

8.2.6 Cramér–Rao Lower Bounds

In this subsection, we present accuracy estimations based on the positioning CRLB for the UMTS cellular network and different measurement types similar to Section 8.1.7. The minimum RX sensitivity of the MT was set to $-103.7\,$dBm (ETSI 2013l). Figure 8.9 demonstrates that *TOA* and *TDOA* measurements in UMTS with a signal bandwidth of 3.84 MHz achieve approximately a 10 times larger positioning accuracy than the corresponding *RSS* measurements. The best performance is provided by the TOARSS and TOA measurements followed by TDOA-RSS and TDOA each having a slightly more degraded performance. The CEPs for the different measurements are summarized in Table 8.7.

Comparing the results between UMTS and GSM (see Section 8.1.7), we see that pure timing based measurements in UMTS perform significant better than RSS measurements in contrast to GSM. For GSM, the RSS measurements perform better than the pure timing based measurements. The maximum transmit power for both systems is 43 dBm. The performance of RSS measurements mainly depends on the received SNR (see Section 2.3.2), that is, the received noise power as the received signal power is the same in both cases. The received noise power is

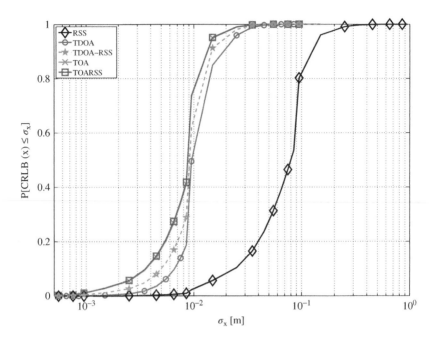

Figure 8.9   CDF of two-dimensional positioning CRLB with different measurements for UMTS

Table 8.7 CEP_x of positioning CRLB for UMTS and different measurements (units are cm)

CEP_x	RSS	TDOA	TDOA-RSS	TOA	TOARSS
$x = 67\%$	9	1.2	1	0.93	0.93
$x = 95\%$	15	2.4	2	1.5	1.5

a function of the receiver bandwidth and the noise power spectral density N_0. Thus, mainly the signal spectra in Figure 8.8 and Figure 8.2(b) determine the positioning accuracy. UMTS has approximately a 10 times larger signal bandwidth than GSM. Thus, we could expect that GSM RSS measurements provide a 10 times higher positioning accuracy than UMTS. Comparing Tables 8.3 and 8.7, we see that the accuracy only improves between a factor of two to three. The smaller than expected gain is mainly due to the multipath propagation of the wireless radio channel affecting the accuracy of the narrowband GSM signal more strongly than that of the wideband UMTS signal.

The UMTS TOA and TDOA measurements are approximately five to nine times more accurate than the corresponding GSM measurements, which is more close to the expected factor of 10. The overall accuracy improvement from GSM to UMTS is between a factor of three to four for the best performing TOARSS measurements.

8.3 3GPP-LTE

To make the 3GPP access technologies highly competitive for the future, 3GPP developed in LTE (ETSI 2012g) and LTE-A (ETSI 2012f) the *evolved UTRA (E-UTRA)* and *evolved UTRAN (E-UTRAN)*. In LTE or LTE-A, the MT is called *UE* as in UMTS and the BS is called *E-UTRAN node B* or *evolved Node B (eNB)*. There are two operation modes for E-UTRA: *FDD* and *TDD*. In general, there exist no separate standard documents for FDD and TDD in E-UTRA in contrast to UMTS, for example, (ETSI 2012h) and (ETSI 2012i). Thus, all the location dependent measurements and positioning methods in 3GPP-LTE are applicable to both FDD and TDD.

Besides the UE and eNB, the E-UTRAN positioning architecture also includes LMUs that can be standalone or part of an eNB similar to GSM or UMTS (see Sections 8.1 and 8.2). Further, E-UTRAN introduces *enhanced serving mobile location centers (E-SMLCs)* (ETS 2013e). E-SMLCs are the central units in LTE for initiating and managing UE based, eNB based, or LMU based positioning procedures. When a UE or network entity requests LCSs, the E-SMLCs will receive the LCS requests and provide an LCS response to the UE or network entity.

Four location techniques have been specified for E-UTRAN (ETS 2013e):

- *Network-assisted GNSS* method can be used in UEs that are equipped with a GNSS receiver, for example, in smartphones.
- The *downlink positioning* method uses *OTDOAs*. These measure the time difference of downlink signals from two different cells at the UE. The UE measures the OTDOAs using assistance data received from the positioning server. The resulting OTDOAs are used to locate the UE in relation to the neighboring eNBs. Since OTDOAs are measured at the UE and involve communication between the UE and eNBs, the downlink OTDOA position-ing method is part of the *LTE Positioning Protocol (LPP)* (ETS 2013d). It also involves *LTE Positioning Protocol A (LPPa)* (ETS 2013c) to collect data from eNBs for supporting downlink OTDOA.
- *E-CID* method, which includes the Cell-ID method, uses additional location dependent mea-surements at the UE or E-UTRAN to improve UE location estimation. The E-CID method employs UE measurements, for example, reference signal received power, reference signal received quality, or UE Rx-Tx time difference, and E-UTRAN measurements, for example, eNB Rx-Tx time difference, timing advance, or angle of arrival (see Section 8.3.2). Similar to UMTS, the E-CID methods for LTE will also include RF pattern matching methods in a future LTE release (3GPP 2013). The E-SMLC receives the UE measurements via the LPP and the eNB measurements via the LPPa protocols (ETS 2013e).
- *Uplink positioning* method employs *uplink relative time of arrival (UL-RTOA)* measure-ments obtained from LMUs that measure the UL-RTOA from received UE reference sig-nals. This network based positioning method involves the LPPa and the *SLm Application Protocol (SLmAP)* to exchange location information and measurements (ETS 2013e). The standardization of this method is currently ongoing and will be completed in a future LTE release (3GPP 2013).

The four positioning methods support different implementations as listed in Table 8.8: UE based, UE-assisted/E-SMLC based, eNB-assisted, and LMU-assisted/E-SMLC based. All four

Table 8.8 UE positioning methods for LTE

Positioning method	UE-based	UE-assisted, E-SMLC-based	eNB-assisted	LMU-assisted, E-SMLC based	SUPL
AGNSS	✓	✓	–	–	✓, UE based and -assisted
Downlink OTDOA	–	✓	–	–	✓, UE-assisted
E-CID	–	✓	✓	–	✓, UE-assisted
Uplink UL-RTOA	–	–	–	✓	–

positioning methods are implemented in the *E-UTRAN control plane* (ETS 2013e). The control plane carries control signals, for example, radio resource control or handover control signals. In addition, some of the methods (see SUPL column of Table 8.8) can be implemented through the *secure user plane location (SUPL)* protocols (ETS 2013e). Here, the necessary data is transmitted in the *user plane* as user data traffic of the network. Thus, SUPL requires minimal modifications to the network. On the other hand, the control plane positioning solutions can support fast, simultaneous localization of multiple UEs, for example in the case of an emergency.

In principle, the AGNSS, OTDOA, E-CID, and UL-RTOA methods for E-UTRAN are the same as for GSM and UMTS (see Sections 8.1.6, 8.2.4, 8.2.3, and 8.1.5). Hence, we do not present further details on the specific methods here. However, we explain in detail the relevant system parameters in Section 8.3.1 and measurements in Section 8.3.2. In Section 3.2, we explain in detail how to measure the propagation time in particular for *OFDM* signals and LTE in the downlink. Additionally, we provide some simulation results for the timing error in Section 8.3.3.

8.3.1 System Parameters

The physical layer of the 3GPP-LTE downlink is basically a coded OFDM system. Different logical channels are encoded using a turbo- or convolutional code. The codebits are interleaved and modulated onto BPSK, QPSK, 16-, or 64-QAM symbols. Data and pilot symbols are multiplexed into OFDM symbols. For details about channel coding, modulation, scrambling, interleaving, precoding, spatial signal processing, and so on, we refer the reader to the respective standard documents (ETSI 2013c,d,e). Here, the focus is on positioning. So, we are interested in those signal parts of 3GPP-LTE that are dedicated to obtaining location information. Subsequently, we briefly introduce the 3GPP-LTE framing structure that embeds the reference signals we are interested in for our investigations. Parameters that influence the positioning accuracy are summarized in Table 8.9 (ETSI 2013b,e,h).

Generally, there are two different types of 3GPP-LTE radio frames, applicable to FDD Type 1 and TDD Type 2. We are focusing on the FDD frame Type 1. Such a radio frame as in Figure 8.10 is 10 ms long and consists of 10 subframes. Each of them consists of two slots containing each seven OFDM symbols. Note, dependent on the cyclic prefix length mode and the subcarrier spacing f_{SC}, that is, the core OFDM symbol length, a subframe may contain six or 12 OFDM symbols. The main difference between FDD frame Type 1 and TDD frame Type 2 is that for FDD all subframes are transmitted either in the downlink or uplink whereas for TDD some subframes are transmitted in the downlink and others in the uplink (ETSI 2013e).

Table 8.9 System parameters of LTE FDD

Parameter	Value
Modulation	BPSK, QPSK, 16-QAM, 64-QAM
FFT length N_{FFT}	2048
Subcarrier spacing	15 kHz
Sampling frequency	30.72 MHz
Sampling duration T_S	32.55 ns
OFDM symbol duration	66.67 µs excluding guard interval
Cyclic Prefix (guard interval)	160 and 144 samples for the first and all other OFDM symbol in a slot
Channel bandwidth	1.4, 3, 5, 10, 15, 20 MHz
Maximum transmitted subcarriers	72, 180, 300, 600, 900, 1200
Occupied bandwidth	1.08, 2.7, 4.5, 9, 13.5, 18 MHz
Slot duration	0.5 ms
Subframe duration	1 ms (2 slots)
Frame duration	10 ms (20 slots)
Maximum MT power	31 or 23 dBm (LTE Band 14)
	23 dBm (all other bands)
Minimum MT power	< 40 dBm
Maximum BS power	no upper limit for wide area BS
	38 dBm medium range BS
	24/20 dBm local area/home BS
Minimum BS power	at least 7.7–20 dB below
	maximum BS power
MT sensitivity	−104.7– − 90 dBm
BS sensitivity	−106.8– − 101.5 dBm wide area BS
	−101.8– − 96.5 dBm medium range BS
	−98.8– − 93.5 dBm local area/home BS

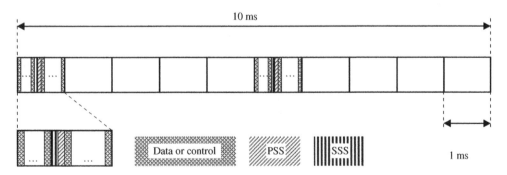

Figure 8.10 3GPP-LTE frame (10 ms), subframes (10 with 1 ms each), data or control (information symbols) and synchronization signals (PSS, SSS)

Besides OFDM symbols that contain data and control information, an LTE frame in the downlink also contains the *PSS* and the *SSS*, which are contained in the seventh and sixth OFDM symbol of the first and sixth subframe. Thus, the PSS and SSS repeat every 5 ms in the LTE frame. 3GPP-LTE supports 1.4, 3, 5, 10, 15, and 20 MHz channel bandwidths, which correspond to 72, 180, 300, 600, 900, and 1200 actively used OFDM subcarriers (see Table 8.9 and ETSI 2013b). In the remainder of this section, we present in more detail the LTE downlink PSS, SSS, *cell-specific reference signals (CRSs)*, and *positioning reference signals (PRSs)* as well as the LTE uplink *sounding reference signals (SRSs)* and *demodulation reference signals (DMRSs)*. All these reference signals are relevant for positioning.

There are three different *PSSs* which are generated from frequency-domain Zadoff–Chu sequences of length 62. These 62 symbols are mapped to the 63 central subcarriers excluding the 0 or *direct current (DC)* subcarrier of the seventh OFDM symbol in subframe 1 and 6 of an LTE Type 1 radio frame (ETSI 2013e). Two of the sequences have the same magnitude and only differ in the phase. The PSS sequences can be used to estimate the integer and fractional frequency offset in the receiver (Berggren and Popovic 2007; Kim et al. 2008). Because there are only three sequences, a bank of matched filters can be used for more precise timing estimation (Tsai et al. 2007).

In total, there are 504 different *SSSs*, which are grouped into 168 unique physical layer cell identity groups, each containing three cell identities. The cell identity within each group is indicated by the PSS, whereas the SSS signal carries the cell identity group number. For the generation of SSS sequences two-length 31 binary m-sequences are interleaved. Each of the resulting 168 sequences are scrambled by one of the three length 31 m-sequences, determined by the PSS signal, that is, the cell identity. The 62 binary frequency-domain samples are mapped to the central subcarriers excluding the DC subcarrier of the sixth OFDM symbol in subframes 1 and 6 of an LTE Type 1 radio frame, that is, they appear prior to the PSS. The SSS sequences are centrally symmetric signals in the time domain. Thus, it is possible to detect the start of the SSS sequences with a *reverse differential correlation* (Berggren and Popovic 2007, see Section 3.2). Although this approach may be less accurate than the matched filter correlator bank for the PSS, it has the benefit of being robust against frequency offsets. As the sequences between subframe 1 and 6 differ, the SSS pilot can be used to detect the frame start.

The CRSs are pilots scattered in time- and frequency over the complete LTE frame (ETSI 2013e). Thus, the CRS occupies a bandwidth of up to 18 MHz with a frequency reuse of 6. In time direction the CRS is contained in every third or fourth OFDM symbol. The CRSs are derived from a length-31 Gold sequence and mapped to QPSK symbols. The CRS can be used for synchronization or channel estimation. The CRSs have a cell specific frequency shift modulo 6. As there are 504 cell identities, 84 CRSs use the same time- and frequency resources.

The PRSs have been newly introduced for the purpose of positioning in 3GPP-LTE Release 9 (ETSI 2013e). The PRS is transmitted in the downlink from the BS to the MT, which measures the OTDOA. The PRSs are transmitted on the separate antenna port 6. The motivation for this is to enable *enhanced IPDL (E-IPDL)* (Qualcomm Europe 2009) to improve hearability of the non-serving BSs. Positioning performance improves with an increasing number of reliable TOA measurements. In dense urban environments, typically, mobiles are able to hear only one or two base stations because of full re-use. To avoid the hearability problem (arising from the near-far effect) of a reuse-1 system, the UTRA standard provides IPDL during which transmission of all channels from a base station ceases. In these periods, the mobile station is able to receive the pilot signal of the neighbor BSs even if the serving BS signal on the same frequency

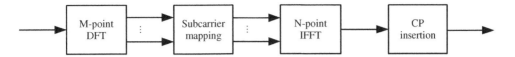

Figure 8.11   3GPP-LTE uplink basic transmission scheme: SC-FDMA, also known as DFTS-OFDM

is very strong. However, in LTE there is no continuous transmission of reference signals or control channels. Hence, the PRSs have been defined to be transmitted in the idle periods. To enable simple receiver processing and to further enhance hearability, a frequency reuse of 6 is defined for the PRS similar to the CRS. The PRS is designed such that it can be used jointly with the CRSs for an acquisition of N_{PRS} consecutive downlink subframes with $N_{\text{PRS}} = 1, 2, 4, 6$. The PRS is punctured by PSS, SSS, and the *physical broadcast channel (PBCH)* as well as by control channels allocated in the first three OFDM symbols of an even-numbered slot. The periodicity of the PRS can be configured to 160, 320, 640 ms, and 1.28 s. Thus, in combination with durations between 1 and 6 ms, the PRS has a duty cycle between 0.08 and 3.75%.

In the uplink, LTE employs as basic modulation scheme *single-carrier FDMA (SC-FDMA)* (ETSI 2013e), which is also known as *DFT-spread OFDM (DFTS-OFDM)* (ETS 2013b). Figure 8.11 depicts the corresponding block diagram. The DMRS and SRS are defined in the uplink, which can be used to obtain location dependent measurements. Both reference signals are derived from Zadoff–Chu sequences (ETSI 2013e).

- *DMRSs* are transmitted in the *PUSCH* and the *physical uplink control channel (PUCCH)* for channel estimation and coherent data demodulation in the eNB. DMRS signals are time multiplexed with the uplink data and occupy the same bandwidth as the uplink data. For normal cyclic prefix, DMRS signals are transmitted in the PUSCH on the fourth SC-FDMA symbol in each subframe.
- *SRSs* are transmitted to provide channel quality information for uplink frequency resource allocation and scheduling algorithms. Hence, SRSs are wideband signals that can occupy in principle the full available uplink transmission bandwidth. Except for special subframes in TDD frame Type 2, the SRS is transmitted in the last symbol of an uplink subframe.

8.3.2   Measurements

The LTE standard (ETS 2013a) specifies the following four distance dependent measurements:

- *Reference signal received power (RSRP)*: This UE measurement is used for cell selection, handover procedures, and the standardized *E-CID* positioning. RSRP is defined as the linear average over the power contributions (in [W]) of the resource elements that carry CRSs within the considered measurement frequency bandwidth (ETSI 2013e). The absolute accuracy of RSRP measurements is ± 6 dB under nominal conditions. The RSRPs are reported from -140 to -44 dBm with 1 dB resolution (ETSI 2013f). From the RSRP and the E-UTRA carrier *received signal strength indicator (RSSI)*, one can derive the *reference signal received quality (RSRQ)* measurement at the UE, which is another distance dependent measurement used in *E-CID* positioning.

- *Reference signal time difference (RSTD)*: This TDOA is measured for *OTDOA* positioning at the UE for at least 16 cells including the serving cell (ETSI 2013f). The RSTD between the neighbor cell j and the reference cell i is defined as $T_{SubframeRxj} - T_{SubframeRxi}$, where $T_{SubframeRxj}$ is the time when the UE receives the start of one subframe from cell j. $T_{SubframeRxi}$ is the time when the UE receives the corresponding start of one subframe from cell i that is closest in time to the subframe received from cell j. The required RSTD accuracy varies depending on the bandwidth of the PRS and can vary between $\pm 5T_S$ and $\pm 15T_S$ for 1200 or 72 subcarriers. RSTD measurements are reported from $-15391T_S$ to $15391T_S$ with $1T_S$ resolution for $|RSTD| \leq 4096T_S$ and $5T_S$ resolution otherwise. Thus, the RSTD measurements provide a spatial resolution of 9.8–49 m.
- *Timing advance* (T_{ADV}): Two T_{ADV} measurements have been defined in LTE (ETS 2013a), which are measured at the eNB. The Type 1 T_{ADV} measures the round-trip delay. The propagation delay is simply $T_{ADV}/2$ between the UE and eNB. The Type 1 T_{ADV} is defined as the time difference T_{ADV} = (eNB Rx–Tx time difference) + (UE Rx–Tx time difference), where the eNB Rx–Tx time difference corresponds to the same UE that reports the UE Rx–Tx time difference. The eNB Rx–Tx time difference is defined as $T_{eNB-RX} - T_{eNB-TX}$, where T_{eNB-RX} denotes the eNB received timing of uplink radio frame #i, defined by the first detected path in time, and T_{eNB-TX} the eNB transmit timing of downlink radio frame #i. Similar, the UE Rx–Tx time difference is defined as $T_{UE-RX} - T_{UE-TX}$, where T_{UE-RX} denotes the UE received timing of downlink radio frame #i, defined by the first detected path in time, and T_{UE-TX} the UE transmit timing of uplink radio frame #i. Figure 8.12 illustrates the relationship between the different measurements to obtain the T_{ADV}. Note that for LTE, the uplink frame #i starts always before the downlink frame #i (ETSI 2013e). Therefore, UE Rx–Tx time difference is always positive and eNB Rx–Tx time difference can be positive or negative. In Figure 8.12, eNB Rx–Tx time difference is negative and thus, T_{ADV} < (UE Rx–Tx time difference). The Type 2 T_{ADV} measures only the eNB Rx–Tx time difference. The round-trip delay can be calculated from the Type 2 T_{ADV} together with the UE Rx–Tx time difference measurement. The T_{ADV} is reported from 0 to $49232T_S$ with resolutions of $2T_S$ for $0 \leq T_{ADV} \leq 4096T_S$ and $8T_S$ for $T_{ADV} > 4096T_S$. Thus, the accuracy of the round-trip delay from T_{ADV} is 9.8 m for $T_{ADV} \leq 40$ km or 39 m for $T_{ADV} > 40$ km.
- *AOA*: Similar to UMTS (see Section 8.2.2), (ETS 2013a) defines the AOA measurement. It estimates the angle of a user with respect to the geographical north reference direction, positive in a counter-clockwise direction. The AOA is determined at the eNB antenna for an UL channel corresponding to this UE and is reported with a resolution of 0.5°.
- *UL-RTOA* $T_{UL-RTOA}$: The UL-RTOA measurement is used in the standardized *UTDOA* positioning method (ETS 2013e). Note that the standardization of the measurement is currently still ongoing. $T_{UL-RTOA}$ is the beginning of subframe i containing SRS received in LMU j, relative to the configurable reference time (3GPP, TSGRAN 2013; ETSI 2013g). The UL-RTOA (3GPP, TSGRAN 2013) should be measured at the RX antenna connector of the LMU node when the LMU has a separate RX antenna or shares an RX antenna with an eNB and the eNB antenna connector when the LMU is integrated in the eNB. The UL-RTOA measurement is reported from $0T_S$ to $9598T_S$ with $2T_S$ resolution, that is, 19.5 m spatial resolution.

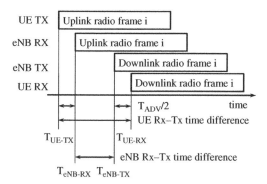

Figure 8.12   3GPP-LTE timing advance Type 1 measurement relationships and uplink-downlink timing relationship

8.3.3   Synchronization

Symbol or frame timing and carrier frequency offset estimation/correction are two tasks that have to be performed prior to any data detection, demodulation, and decoding. Synchronization algorithms are usually based on some knowledge about transmitted signal parts at the receiver, that is, the synchronization or reference signals. Using correlation methods (see Section 3.2), this knowledge can be exploited to estimate the propagation time between the UE and eNB.

For the evaluation of the benefit of location and channel state information, we use a cellular simulation environment setup as shown in Figure 8.13.

BSs are located in a hexagonal grid with a cell radius of of 750 m. We have applied the WINNER C2 Typical Urban multi-link channel model (see Section 8.1.7.2) with a BS TX power of 43 dBm. Three different UE positions–$UE_1 = [100\,m, 0\,m]$, $UE_2 = [500\,m, 0\,m]$, $UE_3 = [750\,m, 0\,m]$–are used for simulations, where UE_3 is located at the cell border. The UE speed is set to 2 m/s. The simulation results that follow show the timing accuracy as *CDFs* over time samples for the *SSS* of 3GPP-LTE. The carrier frequency offset for all the simulations is assumed to be zero.

Figure 8.14 shows the synchronization accuracy for *reverse differential correlation (RDC)* and *cross-correlation (CC)* at different UE locations in the cellular environment. The synchronization accuracy decreases as the UE moves towards the cell border. In particular at the cell edge, the timing performance suffers from interference from adjacent eNBs. As can be seen, CC outperforms RDC. The reason for this is two-fold. First, the correlation length for RDC is half compared to CC. Secondly, RDC is based on correlation of two received signal sections, which doubles the noise.

8.3.4   Cramér–Rao Lower Bounds

In this subsection, we present accuracy estimations based on the positioning CRLB for 3GPP-LTE signals and different measurement types in Figure 8.15 similar to Sections 8.1.7 and 8.2.6. Where not otherwise mentioned, we use the same parameters as in Section 8.1.7.

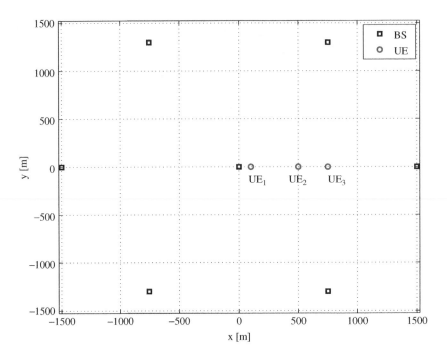

Figure 8.13 Cellular simulation scenario

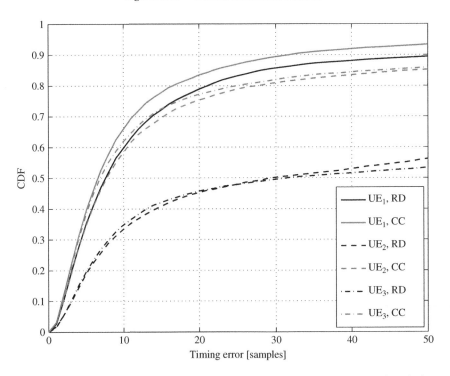

Figure 8.14 SSS synchronization accuracy at different locations of a UE in a cellular wireless communications system for RDC and CC

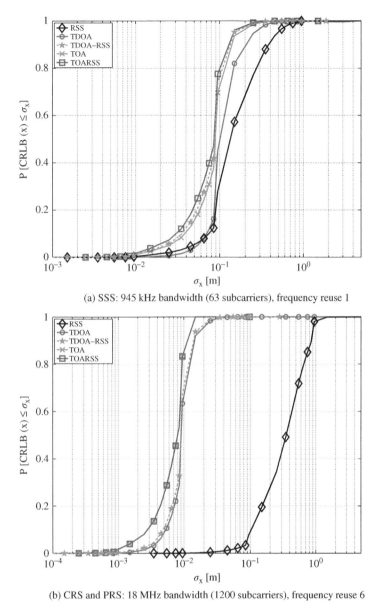

(a) SSS: 945 kHz bandwidth (63 subcarriers), frequency reuse 1

(b) CRS and PRS: 18 MHz bandwidth (1200 subcarriers), frequency reuse 6

Figure 8.15   CDF of two-dimensional positioning CRLB with different measurements for LTE using SSS or CRS and PRS. Note, if the signal power is not boosted as we assume here, the subcarriers without reference signals contain data that cannot be used for positioning

Additionally, we provide an accuracy comparison between the 3GPP systems GSM, UMTS, LTE, and *LTE-A* for TDOA and RSS measurements in Figure 8.16.

The performance of the *LTE SSSs* is plotted in Figure 8.15(a). The minimum RX sensitivity of the UE was set to −94 dBm (ETSI 2013h). Note that the SSS transmit power is not boosted. Here, we assume that the subcarriers outside the central 72 reserved subcarriers excluding

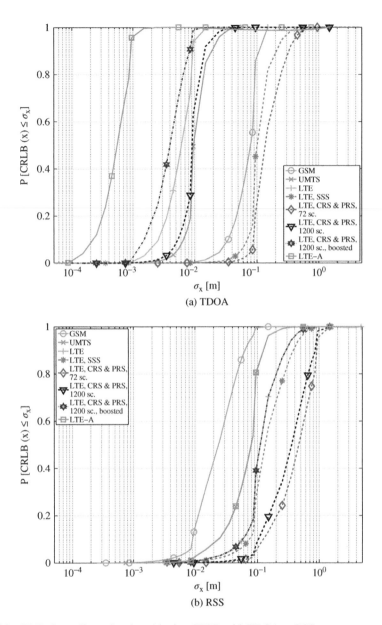

(a) TDOA

(b) RSS

Figure 8.16 CDF of two-dimensional positioning CRLB with TDOA or RSS measurements for different 3GPP systems and signals

the DC subcarrier contain data from users that cannot be used for positioning. As the narrowband SSS only occupies the central 63 subcarriers with a total bandwidth of 945 kHz (see Section 8.3.1), the difference between timing- and RSS based positioning methods is relatively small when comparing it to UMTS. For example, considering the CEPs in Table 8.10

Table 8.10 CEP_x of positioning CRLB for LTE with SSSs or CRSs and PRSs, and for different measurements (units in cm)

Reference signal	CEP_x RSS		TDOA	TDOA-RSS	TOA	TOARSS
SSS	$x = 67\%$	20	13	9.3	9.4	9.2
	$x = 95\%$	49	26	15	18	15
CRS /	$x = 67\%$	50	1	0.95	0.9	0.9
PRS	$x = 95\%$	92	2	1.8	1.3	1.3

Table 8.11 CEP_x of positioning CRLB based on TDOA or RSS measurements for different 3GPP systems and signals (units in cm)

Measurement CEP_x	TDOA		RSS	
	$x = 67\%$	$x = 95\%$	$x = 67\%$	$x = 95\%$
GSM	8.9	13	3.5	8
UMTS	1.2	2.4	9	15
LTE	0.79	1.1	14	38
LTE, SSS	13	26	20	49
LTE, CRS & PRS, 72 sc.	19	41	64	95
LTE, CRS & PRS, 1200 sc.	1	2	50	92
LTE, CRS & PRS, 1200 sc. boosted	0.52	0.9	14	38
LTE-A	0.07	0.1	9	15

for SSSs, the timing- or combined timing-RSS based methods improve by a factor of 1.5 to 3.3 compared to the RSS only method.

Employing the CRS and PRS over 1200 subcarriers (18 MHz occupied bandwidth) with frequency reuse 6 for positioning, Figure 8.15(b) demonstrates an improvement of at least one order of magnitude for the timing based positioning compared to the RSS based positioning. The CEPs in Table 8.10 show that this improvement varies between a factor of 46 and 71. As before, the power of the CRS and PRS are not boosted.

Figure 8.16 and Table 8.11 provide a positioning accuracy comparison between the 3GPP systems GSM, UMTS, LTE, and LTE-A for a UE using either TDOA or RSS measurements. For LTE and LTE-A, we investigate six different signals:

- *LTE*: These curves use 1200 subcarriers (18 MHz bandwidth) containing signals all known to the UE. This signal is basically equivalent to *GSM* and *UMTS* signals, where we assumed that the full channel resource in time and frequency and the full transmit power is assigned to one UE for positioning.
- *LTE, SSS*: These curves use the SSS as described previously and are the same as in Figure 8.15(a). The transmit power of SSS is not boosted.

- *LTE, CRS & PRS, 72 sc.*: These curves use the CRS & PRS on the central 72 subcarriers (1.1 MHz bandwidth) excluding the DC subcarrier with a frequency reuse of 6. Thus, only 12 subcarriers are active in an OFDM symbol containing CRS & PRS. In this case, the CRS & PRS transmit power is not boosted.
- *LTE, CRS & PRS, 1200 sc.*: These curves use the CRS & PRS on the central 1200 subcarriers (18 MHz bandwidth) excluding the DC subcarrier with a frequency reuse of 6. Thus, 200 subcarriers are active in an OFDM symbol containing CRS & PRS. In this case, the CRS & PRS transmit power is not boosted.
- *LTE, CRS & PRS, 1200 sc. boosted*: These curves use the same signal structure as the *LTE, CRS & PRS, 1200 sc.*, but their transmit power is boosted. Here, we assume that the subcarriers without reference signals are empty.
- *LTE-A*: These curves employ carrier aggregation of five adjacent *LTE* carrier signals (ETSI 2013h). Thus, the occupied bandwidth is approximately 98 MHz except for four 2 MHz gaps between adjacent carrier signals. Note, the total transmit power for the *LTE-A* signal has been also increased by a factor of 5. Note that a receiver would have to coherently process the signal with a sampling rate of > 98 MHz to achieve this bound.

From Figure 8.16 and Table 8.11, we draw the following conclusions for TDOA measurements:

- In general, the positioning accuracy increases with increasing bandwidth of the systems and with the same or increasing transmit power.
- The narrowband LTE reference signals, that is, the *SSS* and the *CRS & PRS for 72 subcarriers*, perform worse than narrowband *GSM* even so these signals occupy approximately three times the bandwidth of a GSM signal. However, the transmit power of the narrowband LTE reference signals is not boosted. Thus, their SNR is severely degraded compared to GSM and, they perform worse than GSM.
- The wideband LTE signals (18 MHz bandwidth) perform equal to or better than the UMTS signal for TDOA. The trade-off between increased bandwidth and reduced SNR only improves the positioning accuracy by a factor of 1.5–2.7 for the *LTE* and *LTE, CRS & PRS, 1200 sc. boosted* signals compared to *UMTS*.
- *LTE-A* improves the positioning accuracy by a factor of 10 compared to *LTE* due to the five times larger bandwidth and the $\sqrt{5}$ times larger SNR.

Concerning the RSS measurements, we conclude from Figure 8.16 and Table 8.11:

- In general, the positioning accuracy increases with increasing SNR of the system. Consequently, the SNR increases from *LTE, CRS & PRS, 72sc.* to *LTE, CRS & PRS, 1200 sc.*, to *LTE, SSS*, to *LTE* and *LTE, CRS & PRS, 1200 sc. boosted*, to *UMTS* and *LTE-A*, and to *GSM*.
- Signals with the same SNR result in the same accuracy, for example, the *LTE* and *LTE, CRS & PRS, 1200 sc. boosted* curves.

When comparing the RSS with TDOA measurements, we infer that all signals except for the narrowband GSM signal achieve better positioning accuracies with the TDOA measurements.

8.3.5 Performance Results

For performance results, we simulate a 3GPP-LTE system with the parameters according to Table 8.12. These differ slightly from the standardized 3GPP-LTE system.

The overall transmission bandwidth of the system is 19.8 MHz in this setup. The pilot symbols used for the timing estimation are based on the SSS as specified in (ETSI 2013e) (see Section 8.3.1). For the simulations we assume perfect filtering of the SSS for synchronization.

Figure 8.17 shows the simulated scenario including $N_{BS} = 7$ BSs with three sectors each. The target sector of BS 1 (serving BS) is depicted in gray. In this sector different characteristic MT positions are investigated.

Figure 8.18 depicts the corresponding *RMSEs* of the performed timing estimation with different BSs μ that are defined as

$$\text{RMSE}_\mu = c \sqrt{\text{E}\left\{|\hat{\tau}_{\mu,1} - \tau_{\mu,1}|^2\right\}}.$$

The timing estimates are obtained after averaging over $N_{\text{Frames}} = 100$ frames resulting in an update period of one second. The RMSEs are plotted over the distance from BS 1 for the different MT positions as shown in Figure 8.17. Here, we assume a one-tap channel, that is, we do not face multipath propagation. Totally, 5000 different realizations were simulated for each MT position. As expected, the timing estimation for BS 1 is of high accuracy when the MT

Table 8.12 System and transmission parameters

Parameter	Characteristic/value
Carrier frequency	2 GHz
N_{Used}	1320
N_{FFT}	2048
N_{GI}	128
N_{Symbols}	140
Subcarrier spacing Δf	15 kHz
Cell layout	Hexagonal, inter-site distance of 1500 m
BS transmit power	43 dBm
BS antenna model	3 dB-beamwidth of 70°, 14 dBi
Path loss model	128.1 dB + 37.6 log $_{10}(r_\mu/(1000 \text{ m}))$ dB
Shadow fading model	Log-normal, standard deviation of 8 dB
Multipath model	One-tap or six-tap typical urban
Propagation time	Included
MT antenna model	Omni-directional, 0 dBi
MT noise figure	7 dB
Doppler	According to MT velocity of 3 km/h
Coding type	Conv. code, rate 1/2, with code generator polynomial $(171, 133)_8$ in octal numeral system
Interleaving type	Random
Data mapping	16-QAM, gray mapping
Channel estimation	Perfect in frequency domain

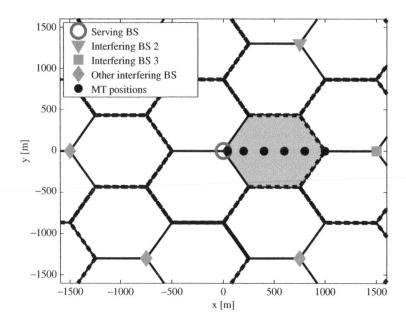

Figure 8.17   Cellular network structure

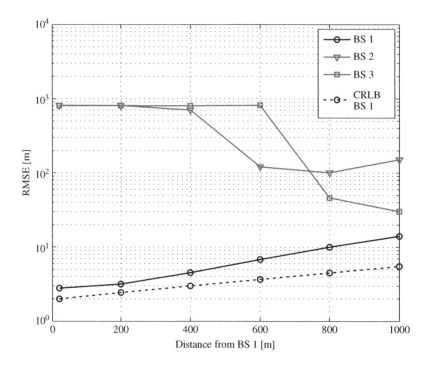

Figure 8.18   RMSE of timing estimation

is in the inner cell. Moving to the cell edge, the performance becomes worse. For comparison, also the corresponding CRLB for BS 1 is plotted here. The timing estimates are close to this CRLB in the inner cell, since the assumptions for the derived estimator are nearly fulfilled here (no multipath, no interference). For the timing estimates of BS 2 and BS 3 we observe that only at the cell edge a good quality can be provided. Note that due to geometrical *a priori* knowledge (inter-site distance is known at the MT) the search space for the timing estimation in (3.22) can be reduced, and hence, also the maximum error is upper bounded. This explains the flat curves for BS 2 and BS 3 in the inner cell. Nevertheless, in these situations the quality of the timing estimates is insufficient for precise positioning.

Finally, in Figure 8.19 the performance after the position estimation process is plotted. For positioning we employ the strongest $N_{\text{BS, Used}} = 3$ BSs. The initial value for the iterative Gauss–Newton algorithm (see Section 4.3.2.1) is the mean value of the positions of the involved BSs, that is,

$$x^{(0)} = \frac{1}{N_{\text{BS, Used}}} \sum_{\mu=1}^{N_{\text{BS, Used}}} x_\mu.$$

As performance criterion we use the CDF (see Section 4.5.2).The CDF was averaged over several MT positions in the scenario and noise realizations.

We observe that in case of the one-tap channel (no multipath) we achieve a $\text{CEP}_{67\%} < 247$ m. In case of multipath propagation, the overall accuracy decreases. For instance, the $\text{CEP}_{67\%} < 281$ m. Thus, the error is only increased by about 14% due to multipath propagation. The reason

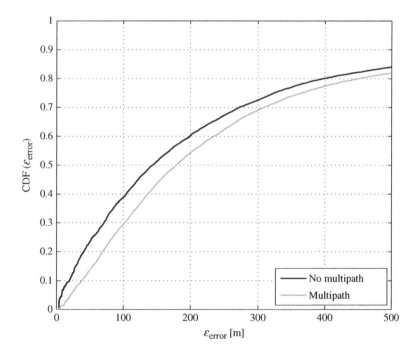

Figure 8.19   CDF of location estimation error

for this behavior is that in several situations the position cannot be determined with sufficient accuracy anyhow−in these situations additional multipath propagation will not have a high influence. Note that the multipath effects could be reduced by a multipath mitigation scheme as mentioned already in Chapter 7. When comparing the CRLBs in Figure 8.16 with the RMSE in Figure 8.19, we see that the positioning accuracy is decreased due to interfering signals by more than two orders of magnitude. Thus, clearly interference avoidance techniques, for example, E-IPDL (Qualcomm Europe 2009), or interference mitigation techniques, for example, interference cancelation (Mensing et al. 2009), are necessary to get close to the theoretical positioning performance of LTE.

8.4 Other Wide and Medium Range Systems

In this section, we present two more commercial wide and medium range communications systems that partially use OFDM, which are currently or in the near future relevant for positioning: WiMAX and WLAN.

8.4.1 WiMAX

Like 3GPP-LTE, mobile WiMAX is based on OFDM modulation (Etemad 2008; Wang et al. 2008a) and provides flexibility in terms of the basic OFDM parameters, which define, for example, symbol length or occupied bandwidth. WiMAX is a subset of the (IEEE 802.16 2012) standard. Table 8.13 shows typical choices for these parameters, supporting different channel bandwidths.

WiMAX uses a well established framing structure, where each frame in the downlink consists of a preamble OFDM symbol, followed by control information symbols and the payload symbols. The preamble symbol is known at the receiver and, therefore, can be used for synchronization and channel estimation. A set of subcarriers of the first OFDM symbol in the uplink is foreseen for ranging purposes, which allows the base station to calculate a timing advance parameter for the respective MT. This ensures that the OFDM signal of all the MTs arrive within the OFDM timing window at the base station, regardless of their distance from the base station. This timing advance is therefore a direct measure of the distance between BS and MT. The principle behind this inherent ranging method is the *RTTOA* method. In addition, the WiMAX standard specifies *TDOA* for localization of MTs in the uplink with *UTDOA* and in the downlink with *EOTD* (Venkatachalam et al. 2009). Bandwidths and OFDM symbol timing

Table 8.13 Typical WiMAX OFDM parameters

FFT length	Subcarrier spacing	OFDM symbol length	Used carriers	Bandwidth (MHz)	System sampling time (ns)
128	10.94 kHz	91.43 μs	114	1.25	714.0
256			228	2.5	357.0
512			457	5	178.5
1024			914	10	89.3
2048			1828	20	44.6

parameters are quite similar compared to 3GPP-LTE. The synchronization performance–and therefore the ranging performance–can therefore be expected to be in the same order of magnitude like those of 3GPP-LTE.

Similar to LTE and LTE-A, the (IEEE 802.16.1 2012) standard, of which a subset may become the *WiMAX-Advanced* system, specifies the Advanced Air Interface of (IEEE 802.16 2012). (IEEE 802.16.1 2012) employs besides basic *location based services (LBS)* support as in (IEEE 802.16 2012) also enhanced LBS support through a downlink LBS zone with a cell specific preamble to improve hearability; thus, enhancing downlink positioning (Wang et al. 2008b).

8.4.2   WLAN[1]

WLAN, also known as WiFi, is nowadays commonly available in public and office buildings, which makes it one of the most promising indoor positioning methods in the near future, at least for inexpensive consumer products. According to (Hiertz et al. 2010), five major *physical layer (PHY)* implementations of WLAN exist that are widely used based on *direct-sequence spread spectrum (DSSS)* or *OFDM* modulation techniques: the original 802.11 DSSS PHY, the 802.11b *high rate DSSS (HR/DSSS)* PHY, the 802.11g *extended rate PHY (ERP)* with DSSS or *complementary code keying (CCK)*, the 802.11a OFDM PHY, 802.11g ERP-OFDM PHY, and the 802.11n *high throughput (HT)* PHY. Table 8.14 summarize the positioning relevant parameters of these PHYs.

Positioning is supported by the current WLAN standard (IEEE 802.11 2012) through amendments (IEEE 802.11k 2008; IEEE 802.11u 2011; IEEE 802.11v 2011) in contrast to the previous standard (IEEE 802.11 2007). The location of a mobile device may be determined in multiple ways including *RSS*, *TOA*, and *TDOA*. Note that the timing based method specified in (IEEE 802.11 2012) is using *RTTOAs*. The timing measurements to obtain the RTTOA are reported in 10 ns intervals, that is, a spatial resolution of 3 m. The timing error Δt can be $\Delta t \leq \pm n_{err} \cdot 10$ ns with $n_{err} = 1, \ldots, 254$, $\Delta t \geq \pm 2.55 \,\mu$s for $n_{err} = 255$, or unknown ($n_{err} = 0$).

WLAN positioning methods based on either RSS or TOA measurements are reported in the literature (GRAMMAR 2009a). Examples of RSS based WLAN positioning can be found in (Bahl and Padmanabhan 2000; Prasithsangaree et al. 2002; Roos et al. 2002; Smailagic and Kogan 2002; Wallbaum and Wasch 2004) and examples of TOA based WLAN positioning in (Ciurana et al. 2007; Golden and Bateman 2007; Izquierdo et al. 2006). Furthermore, (Prieto et al. 2009) present experimental results on WLAN positioning using RTTOA measurements in accordance to (IEEE 802.11v 2011). The authors report a $CEP_{67\%} \approx 5$ m for an indoor scenario.

Concerning commercial products that use WLAN for LCS, (AeroScout 2013) uses RSS and TDOA measurements in their WLAN products obtained by a proprietary location engine. However, currently no products exist that use RTTOA measurements as specified in (IEEE 802.11 2012). Thus, most WLAN products use RSS measurements for LCS. The MT can obtain the RSS measurement by passive scanning of WLAN beacon frames, which WLAN *access points (AP)* emit periodically. In addition, in many mobile devices, such as smartphones, tablet computers, and laptop computers, RSS measurements are easily available through *application programming interfaces (APIs)* of their standard WLAN services.

---

[1] Source: (GRAMMAR 2009a), Section 1.3. 'Positioning with WLAN'. Reproduced with permission of Helena Leppäkoski.

Table 8.14 System parameters of WLAN 802.11 standards

Parameter	Value			
Standard	802.11	802.11b / 802.11g	802.11a / 802.11g	802.11n
Year	1997	1999 / 2003	1999 / 2003	2009
PHY specification	DSSS	HR/DSSS / ERP-DSSS/CCK	OFDM / ERP-OFDM	HT
Modulation technique	DSSS	DSSS	OFDM	OFDM
ISM Band [GHz]	2.4	2.4	5, 2.4	2.4, 5
Symbol / Subcarrier Modulation	Differential BPSK, Differential QPSK, 8-chip CCK		BPSK, QPSK,16-QAM 64-QAM	
Channel Bandwidth [MHz]	20	20	5, 10, 20	5, 10, 20 40
Minimum AP power	1 mW	–	–	–
Maximum AP power	typically 100 mW for all types			
MT RX sensitivity [dBm]	–80	–76	–88, –85, –82 for 5, 10, 20 MHz	–79 for 40 MHz
MT RX maximum input [dBm]	–4	–10	–30 for 5 GHz –20 for 2.4 GHz	
Spreading code	11-chip Barker code 8-chip CCK		–	–
Chip rate	11 MHz		–	–
FFT length	–	–	64 for 5, 10, 20 MHz, 128 for 40 MHz	
Subcarrier spacing [kHz]	–	–	78.125, 156.25 for 5, 10 MHz, 312.5 for 20, 40 MHz	
OFDM symbol duration [μs]	–	–	12.8, 6.4 for 5, 10 MHz, 3.2 for 20, 40 MHz	
Used carriers	–	–	52 for 5, 10, 20 MHz, 56, 114 for 20, 40 MHz HT	
Occupied bandwidth [MHz]	–	–	4.0625, 8.125, 16.25, for 5, 10, 20, 17.5, 35.625 for 20, 40 HT	
Sampling time [ns]	–	–	200, 100, 50, 25 for 5, 10, 20, 40 MHz	

Hence, we consider in the remainder of this section only WLAN RSS based positioning. The algorithms used in WLAN RSS based positioning can be divided into three main categories: Cell-ID based methods, trilateration, and fingerprinting.

8.4.2.1 Cell-ID

In the Cell-ID method, the MT scans the available WLAN channels (GRAMMAR 2009a). As its position estimate, it reports the position of the AP from which it received the strongest signal. In the Cell-ID method, a MT needs prior information about the locations of APs and their unique *media access control (MAC)* addresses. The method is applicable in scenarios where high accuracy is not required or where position technology is not the main focus. The obtainable accuracy is dominated by two factors: the distances between APs in the WLAN network, generally introducing considerable granularity of the positioning, and noise in RSS measurements caused by environment. Nowadays, when most WLAN base stations offer more than one channel to be detected on WLAN scans, the effect of measurement noise can be decreased using RSS based weighting and counting to increase the reliability of the positioning results of the Cell-ID method (di Flora and Hermersdorf 2008; Hermersdorf 2006). The Cell-ID method is used in commercial LCS products, for example, from Apple and Google (Apple 2013; Google 2013).

8.4.2.2 Trilateration

In trilateration systems, path loss models of radio signals are used to translate RSS measurements to distances between the receiver and APs (GRAMMAR 2009a; Smailagic and Kogan 2002; Wallbaum and Wasch 2004). As in Cell-ID based positioning methods, the MT needs prior information about the MAC addresses and locations of APs. In indoor environments, multipath, and attenuation caused by walls, other structures, and even people complicate the modeling of signal propagation. This makes the simple path loss models too inaccurate in many real life situations. To overcome this problem, the performance of triangulation can be enhanced using other models, such as pattern matching (Smailagic and Kogan 2002) and probabilistic filtering approach (Wallbaum and Wasch 2004).

8.4.2.3 Fingerprinting

Fingerprinting approaches are based on experimental models that relate the measured RSS values to the measurement position (GRAMMAR 2009a). These experimental models, also called radio maps, are based on off-line collected data from several locations that sufficiently cover the area of the radio map. Fingerprinting algorithms are considered to be more robust against signal propagation errors as they actually make use of location dependent error characteristics of radio signals. The procedure for radio map creation is often called calibration or training, referring to calibration or training of the experimental model and the required data is called calibration or training data. The locations where the calibration data is collected are called *calibration points*. In the estimation phase, new measurement vectors are related with the information stored in the radio map. A known disadvantage in fingerprinting approaches is

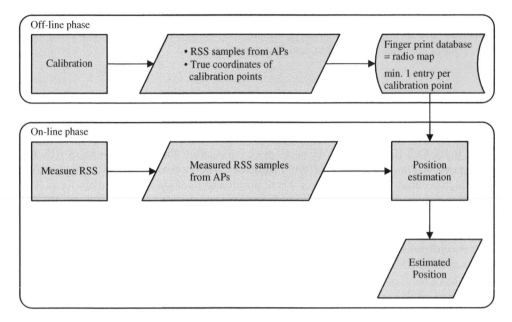

Figure 8.20   Two processing phases of fingerprinting based positioning. Source: (GRAMMAR 2009a), Section 1.3. 'Positioning with WLAN'. Reproduced with permission of Helena Leppäkoski

the fact that the collection of calibration data is laborious and time consuming. The two phases of fingerprinting based position estimation are shown in Figure 8.20.

8.5   Short Range

In this section, we present the Bluetooth, ZigBee, UWB, RFID, and NFC short range communications systems and how they are used for positioning.

8.5.1   Bluetooth

Bluetooth is a wireless communications standard for exchanging data over short distances in the 2.4 GHz ISM band (Bluetooth SIG 2010). It supports over the air data rates from 1–3 Mbit/s. Bluetooth uses a *Gaussian frequency-shift keying (GFSK)* modulation with bandwidth-time product $BT = 0.5$ in the *basic rate (BR)* mode or in the *low energy (LE)* mode and one bit per frequency shift. In the *enhanced data rate (EDR)* mode, the two or three bits are modulated either with $\pi/4$-*differential PSK (DPSK)* or 8-DPSK. Bluetooth uses *frequency-hopping* as spread spectrum technology on 79 bands with 1 MHz bandwidth between 2402 and 2480 MHz for the BR and EDR modes and 40 bands with 2 MHz bandwidth for the LE mode.

Bluetooth is a packet based protocol forming a *piconet* with a *master-slave structure* to form *wireless personal area networks (WPANs)* (Bluetooth SIG 2010). The master shares the channel through *TDD* with up to seven active slaves. Table 8.15 summarizes the Bluetooth

Table 8.15 Bluetooth power classes and transmit ranges

Power class	Maximum output power [dBm]	Range [m]
1	20	100
2	4	10
3	0	1
LE	10	17

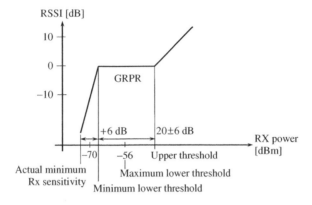

Figure 8.21 Example of Bluetooth RSSI parameter versus RX power. Note that the *GRPR* is defined as 0 dB RSSI

device power classes, their maximum transmit powers and ranges (Bluetooth 2013). Note, the range for the LE power class has been estimated through interpolation. The minimum RX sensitivity shall be below −70 dBm.

The only two distance dependant measurements in Bluetooth are the *link quality (LQ)* and the *RSSI* parameters. The LQ parameter ranges from 0–255. A higher LQ parameter indicates a better link quality. The calculation of the LQ parameter is not standardized and thus, vendor dependent. For many Bluetooth devices, it is derived from the average bit-error rate measured at the receiver (Hossain and Soh 2007).

The RSSI parameter returns the difference between the measured RSSI and the limits of the *Golden Received Power Range (GRPR)* (Bluetooth SIG 2010). An example of the relation between the RSSI, GRPR, and RX power is depicted in Figure 8.21. Clearly, the RSSI contains distance dependent information, but is dependent on the specific device implementation of the GRPR. Moreover, the distance uncertainty within the GRPR is high. Thus, Bluetooth RSSI based positioning can only provide limited positioning accuracy. The RSSI can be obtained with the *Read RSSI Command*. In this case, it is measured with an accuracy of ±6 dB and reported in a range from −128–127 dB for BR and EDR modes and −127–20 dB for the LE mode. Additionally, the RSSI can be obtained from several devices with the *Inquiry Result with RSSI* event in the range of −127–20 dB.

Indoor Positioning with Bluetooth has been widely investigated in the literature. The three main positioning approaches are:

- Proximity based positioning, that is, *Cell-ID* based positioning, is considered in (Forno et al. 2005; Hallberg et al. 2003; Li et al. 2012) as the RSSI measurements are often unreliable and inaccurate.
- *Distance based positioning with RSS or RSSI* and propagation models use an RX power measurement (Hossain and Soh 2007; Kotanen et al. 2003; Zhou and Pollard 2006). Whereas (Hossain and Soh 2007) directly obtains the RSS measurement from the Bluetooth device, (Kotanen et al. 2003; Zhou and Pollard 2006) obtain the RSS indirectly through a mapping of the Bluetooth RSSI to the RSS (see Figure 8.21). From the RSS all three papers then estimate with a propagation model (path loss model) the distance between the transmitting and receiving Bluetooth device.
- *Fingerprinting* positioning with RSSI or *signal strength difference* are investigated in (King et al. 2009; Mahtab Hossain et al. 2013; Wendlandt et al. 2005; Zhang et al. 2013). As in Section 8.4.2.3, these methods use two steps: An off-line training phase to learn and store the fingerprints and an online phase to estimate the position from previously stored fingerprints.

Moreover several papers consider hybrid Bluetooth and WLAN based positioning, for example, (Baniukevic et al. 2011; Mahtab Hossain et al. 2013; Zhu et al. 2012).

Several commercial products (infsoft GmbH 2013; POLE STAR 2013; Teldio 2013) exist that use Bluetooth together with other technologies for indoor positioning. Last but not least, the In-Location Alliance (In-Location Alliance 2013), which was recently founded by industry, considers Bluetooth and WLAN as two enabling technologies to drive innovation and market adoption of high accuracy indoor positioning and related services.

8.5.2   ZigBee

ZigBee is a wireless communications standard standard for low-cost, low-power, and low-rate *WSNs* and *WPANs* as well as *WLANs* (ZigBee 2007; ZigBee Alliance 2013). It addresses diverse applications such as building and home automation, retail and telecommunications services, smart energy, and health care. A ZigBee network supports star, tree, and mesh topologies (ZigBee 2007). The PHY and MAC layer of ZigBee implement the (IEEE 802.15.4 2003) standard. For the channel access, ZigBee supports unslotted *carrier sense multiple access with collision avoidance (CSMA/CA)* and optionally a superframe with beacon, contention access period, and contention free period slots. For the contention access period, devices use slotted CSMA/CA. In the contention free period, devices transmit in guaranteed time slots, that is, in a TDMA scheme for low latency applications.

The ZigBee PHY system parameters according to (IEEE 802.15.4 2003) are summarized in Table 8.16. ZigBee supports energy detection and link quality indication measurements for the network management.

Similar to Bluetooth (see Section 8.5.1), ZigBee location systems are widely studied in the literature and there exist also commercial location systems:

- Localization using the *RSS* distance dependance or RSS *fingerprints* is researched for instance in (Cherntanomwong and Suroso 2011; Fang et al. 2012). The authors report average position accuracies of approximately 1 m from experiments.

Table 8.16 PHY system parameters of ZigBee

Parameter	Value		
PHY [MHz]	868	915	2450
Frequency band [MHz]	868–868.6	902–928	2400–2483.5
Coverage	EU	US	globaly
Number of channels	1	10	16
Channel spacing [MHz]	–	2	5
Spreading technique	DSSS	DSSS	DSSS
Chip rate [kchip/s]	300	600	2000
Pulse shape	raised cosine, roll-off factor 1	raised cosine, roll-off factor 1	half-sine
Occupied bandwidth [MHz]	0.6	1.2	2
Symbol modulation	BPSK	BPSK	Offset-QPSK
Bit rate [kb/s]	20	40	250
Symbol rate [ksymbols/s]	20	40	62.5
Symbols	Binary	Binary	16-ary orthogonal
Maximum TX power [dBm]	≥ -3	≥ -3	≥ -3
RX sensitivity [dBm]	-92	-92	-85
RX maximum input [dBm]	≥ -20	≥ -20	≥ -20

- In contrast to Bluetooth (see Section 8.5.1) (Bedford and Kennedy 2012; Maheshwari and Kemp 2011; Pichler et al. 2009) investigate *TOA* or *TDOA* based positioning. The reported positioning accuracies of (Bedford and Kennedy 2012; Maheshwari and Kemp 2011) vary from 2–10 m. These reduced accuracies are mainly due to the narrowband ZigBee signal and multipath propagation. In contrast, (Pichler et al. 2009) reports accuracies of approximately 1 m using a modified PHY with a bandwidth of 80 MHz.
- Examples of commercial products for ZigBee location systems are the JN5148-001 wireless micro-controller (NXP Semiconductors 2013), which measures both RSS and time-of-flight, the AT86RF233 radio transceiver (Atmel 2013), that measures time-of-flight, and the CC2431 system-on-chip (Texas Instruments 2013), which measures only RSS.

8.5.3 Ultra-Wideband – UWB

UWB communications systems are short-range radio communications systems that spread the signal energy over a very large frequency range (FCC 2002; ITU-R 2006). Thus, UWB systems may overlap several other radio communications technologies and services. Typical UWB technology intentionally radiates signals with at least 500 MHz bandwidth or a fractional bandwidth greater than 0.2. Both bandwidths are measured at the −10 dB bandwidth. The fractional bandwidth is the fraction between the −10 dB bandwidth and the center frequency. Clearly, the large bandwidth of UWB systems promises precision timing based ranging.

There exist several proprietary and standardized UWB communications technologies that can determine the location of UWB devices. In the following we focus on the standardized (IEEE 802.15.4a 2007) and (ECMA 2008; WiMedia 2009b) in more detail:

- (IEEE 802.15.4a 2007) defines an alternative PHY for data communication with precision ranging based on *impulse radio (IR) UWB* as an amendment to the IEEE 802.15.4-2006 standard. This amendment has been integrated into the current (IEEE 802.15.4 2011) standard. The PHY parameters are summarized in Table 8.17.

The precision ranging should be accurate to one meter or better. To achieve this, UWB ranging devices measure the *RTTOA* with two-way ranging (see Section 2.1.3). Furthermore, the ranging devices can use symmetric double-sided two-way ranging to reduce the ranging

Table 8.17   (IEEE 802.15.4a 2007) UWB PHY system parameters

Parameter	Value		
Frequency band [GHz]	0.2496–0.7496	3.1–4.8	6–10.6
Number of channels	1	4	11
Channel bandwidth [MHz]	499.2,1081.6,1331.2,1354.97		
Modulation technique	IR-UWB		
Pulse shape	root raised cosine, roll-off factor 0.6		
Pulse duration [ns]	2, 0.92, 0.75, 0.74		
Data modulation	combined burst position modulation and BPSK		
Data rates	110 kb/s, 851 kb/s (mandatory),		
	6.81 Mb/s, 27.24 Mb/s		
Maximum TX power [dBm/MHz]	regulation dependent, e.g., FCC maximum −41.3		
RX sensitivity [dBm]	–		
RX maximum input [dBm/MHz]	≥ −45		
Synchronization preamble of PHY frame			
Code alphabet	ternary −1,0,1		
Code length	31	31	127 (optional)
Spreading	Dirac function with $L − 1$ zeros		
Spreading length L	16	64	127
Chips per symbol	496	1984	508
Peak pulse repetition frequency [MHz]	31.20	7.80	124.80
Symbol duration [ns]	993.59	3974.36	1017.63
Number of symbols	16 short, 64 default, 1024 medium, 4096 long		
Sync preamble duration			
Short [µs]	15.9	63.6	16.3
Default [µs]	63.6	254.4	65.1
Medium [µs]	1017.4	4069.7	1042.1
Long [µs]	4069.7	–	4168.2
Ranging			
Counter	32 bit unsigned integer		
Counter resolution	15.65 ps ≙ 4.7 mm		
Maximum counter value	67.2 ms		
Confidence level(%)	None, 20, 55, 75, 85, 92, 97, 99		
Confidence interval	100 ps, 300 ps, 1 ns, 3 ns		
Confidence interval scaling factor	0.5, 1, 2, 4		

errors due to clock drifts between two devices. Besides RTTOA estimates, (IEEE 802.15.4a 2007) also considers TOA and TDOA estimates for positioning. In this case, the anchor nodes with known positions need to be time synchronized.

Similar to ZigBee (IEEE 802.15.4 2011), (IEEE 802.15.4a 2007) supports energy detection and link quality indication location-dependent measurements for the network management. Additionally, the energy detection measurements can be computed on a fraction of the total channel bandwidth for the purpose of active detect and avoid procedures to avoid harmful interference with other radio systems.

- The WiMedia Alliance and ECMA International have standardized the *Multi-Band (MB)-OFDM UWB* system (ECMA 2008; WiMedia 2009b). It is a high-speed UWB system that offers an unsurpassed combination of high data throughput rates and low energy consumption (WiMedia 2013). With regulatory approval in major markets worldwide, this technology has gained broad industry momentum as evidenced by its selection for Wireless USB (Hewlett-Packard et al. 2010) and high-speed Bluetooth.

The PHY layer parameters are summarized in Table 8.18. Note, the frequency hopping sequence is defined through time-frequency codes. WiMedia UWB devices can measure the link quality indicator, which is an estimate of the SNR, and the RSSI, which is the energy received at the antenna of the device. The WiMedia UWB PHY also specifies optional ranging and location awareness in addition to high data rate transmission (ECMA 2008; WiMedia 2009b). The ranging accuracy shall be ± 60 cm or better (see Table 8.18). As distance dependent measurements, ranging devices can estimate the *RTTOA* from one or multiple two-way ranging measurements (WiMedia 2009a).

Researchers have extensively studied positioning with UWB systems see (Dardari et al. 2009; Fontana 2004; Gezici et al. 2005; Ingram et al. 2004; Lee and Scholtz 2002; Wymeersch et al. 2009) and references therein. Due to the large bandwidth and the support for precision ranging (IEEE 802.15.4a 2007), most authors focus on TOA, TDOA, or RTTOA ranging methods. However, some also consider the fusion of timing based measurements with RSS measurements to improve location accuracy in short-range indoor environments (Gezici et al. 2005). Most of the researchers consider IR-UWB and the (IEEE 802.15.4a 2007) system for precision ranging. Nevertheless, a few (Cardinali et al. 2006; Xu et al. 2008) also consider MB-OFDM according to (ECMA 2008; WiMedia 2009b).

Several commercial UWB system exists for positioning and LBSs (DecaWave 2013; TimeDomain 2013; UbiSense 2013; ZEBRA 2013). To the best of our knowledge, only (DecaWave 2013) implements the ranging according to (IEEE 802.15.4a 2007). None of the (WiMedia 2009b) UWB systems currently implement ranging. (ZEBRA 2013) implements the novel (IEEE 802.15.4f 2012) standard for active RFIDs with UWB radio technology.

8.5.4 Radio-Frequency Identification and Near Feld Communication– RFID and NFC

Another wireless technology that has been in use for positioning is RFID (ETSI 2006). In conventional RFID, tags are attached to an object or person. By passing a fixed reader with known location, the location of the object or person can be determined (Miller et al. 2006).

Table 8.18   WiMedia UWB PHY system parameters

Parameter	Value
Standard	(ECMA 2008; WiMedia 2009b)
Frequency band [GHz]	3.1–10.6
Number of channels	14
Channel bandwidth [MHz]	528
Channel groups	6
Channels per group	3 for group 1–4, 6 and 2 for group 5
Modulation technique	MB-OFDM, frequency hopping within channel group
Sampling frequency [MHz]	528
FFT length	128
Subcarrier spacing [MHz]	4.125
OFDM symbol duration without /	242.42
with guard interval [ns]	312.5
Frequency hopping rate /	3.2
OFDM symbol rate [MHz]	
Used carriers subcarriers	110
Used bandwidth [MHz]	453.75
Maximum TX power [dBm/MHz]	regulation dependent, e.g. FCC maximum −41.3
RX sensitivity [dBm]	−80.8, −70.4, −63, 5 for 53.3, 480, 1024 Mb/s
	Ranging
Accuracy and precision	≤ 60 cm
	frequency [MHz] 528 1056 2112 4224
Ranging timer resolution	time [ps] 1894 947.0 473.5 236.7
	distance [cm] 56.8 28.4 14.2 7.10
Ranging timer maximum	62.1 μs, 3.97 ms, 1.02 s for 15–18, 21–24, 29–32 bits per active timer

This proximity based approach is suitable for a large number of objects or persons and low location accuracy, for example, in a warehouse.

More recently, reverse RFID, where the tags are installed at fixed known locations and the moving objects or persons carry readers, is considered for location determination (Miller et al. 2006). This approach is suitable for a small number of moving objects or persons and good location accuracy.

Two types of tags can be distinguished. The first type are passive tags, which are activated by the RF power radiated from the reader. The second type are active tags, which contain a battery and can transmit signals independent of a reader. Depending on this, the RFID reader can just determine the proximity of the tags or measure the RSSs to determine the distance of each tag, for example, via fingerprinting (Wendlandt et al. 2007).

For RFID, several different technologies exist at the following frequencies suitable for location determination (ETSI 2006; Miller et al. 2006): 433, 860–960 MHz, 2.45 GHz. The advantage of these technologies compared to lower frequency RFID tags are that they can use small antennas and they operate in the far field having a power decay proportional to one over the distance squared. For example, (Wendlandt et al. 2007) uses 433 MHz active RFID tags with RSS based fingerprinting whereas (Miller et al. 2006) also considers 900 MHz passive RFID tags for proximity based localization. For passive RFID tags, (Alippi et al. 2006) reports accuracies of 0.6 m using a Bayesian statistical method based on measurements from multiple RFID readers.

RFID technology has been standardized by several organizations. For instance, the ISO/IEC standards (ISO/IEC 18000 2008-2013) provide RFID specifications for item management in the 135 kHz, 13.56 MHz, 433 MHz, 860 − 960 MHz, and 2.45 GHz RF bands. ISO/IEC also considers RFID smart cards with maximum read ranges of 1 − 1.5 m (ISO/IEC 15693 2008-2010) or just a few centimeters (ISO/IEC 14443 2008-2012) at the 13.56 MHz HF band. The EPCglobal joint venture also specified RFID standards (EPC HF RFID 2011; EPC UHF RFID 2008) in the 13.56 MHz and 860 − 960 MHz bands to support the *electronic product code (EPC)*. The EPC is a unique number that identifies a specific item in a supply chain. The IEEE has recently standardized the PHY layer for active RFIDs (IEEE 802.15.4f 2012). This active RFID technology can be used in a wide range of applications requiring various combinations of low cost, low energy consumption, multi-year battery life, reliable communications, precision location, and reader options. An example of a commercial product on this standard is (ZEBRA 2013).

NFC is a new RFID technology in smartphones to enable simple paired communication between two devices by bringing them in close proximity. NFC is promoted by (NFC Forum 2013) and standardized by (ISO/IEC 18092 2013; ISO/IEC 21481 2012), (ECMA 2013a,b), and (ETSI 2003, 2004). NFC provides low data rates of 106, 212, and 424 kbps in the 13.56 MHz HF band (NFC Forum 2013). NFC uses inductive coupling, where loosely coupled inductive circuits share power and data over a distance of a few centimeters. NFC can be used to bootstrap Bluetooth or WLAN communication, for example, to exchange photos. An NFC-enabled device can operate as reader or writer as well as in peer-to-peer mode. NFC also incorporates (ISO/IEC 14443 2008-2012) standard to define NFC tags, which are typically passive and store data that is read by an NFC-enabled device.

In the literature, several researchers consider proximity based localization and navigation with NFC tags (Bittins and Sieck 2012; Hammadi et al. 2012; Ozdenizci et al. 2011) for indoor positioning. For instance, (Bittins and Sieck 2012) reports indoor positioning experiments with a Google Nexus S smartphone employing a particle filter with 200 particles. The particle filter fuses measurements of an inertial navigation sensor, a compass and NFC tags. The NFC tags, which have a range of 10 cm, provide absolute location information to the particle filter of 10 cm accuracy.

8.6   Standardization

In the previous sections, we have introduced several standards relevant for positioning with wireless communications systems, which are:

- the 3GPP standards GSM, UMTS, and LTE (ETS 2013e; ETSI 2012a,j),
- WiMAX (IEEE 802.16 2012; IEEE 802.16.1 2012),
- WLAN (IEEE 802.11 2012),
- Bluetooth (Bluetooth SIG 2010),
- ZigBee (IEEE 802.15.4 2003; ZigBee 2007),
- UWB (ECMA 2008; IEEE 802.15.4a 2007; WiMedia 2009b), and

- RFID and NFC (EPC HF RFID 2011; EPC UHF RFID 2008; IEEE 802.15.4f 2012; ISO/IEC 14443 2008-2012; ISO/IEC 15693 2008-2010; ISO/IEC 18000 2008-2013) and (ECMA 2013a,b; ISO/IEC 18092 2013; ISO/IEC 21481 2012).

In addition to these standards, there exist many other standardized wireless technologies and standards that are relevant or may become relevant in the future for wireless positioning. The following list just mentions some of these, which have not been addressed in the previous sections:

- The 3GPP2 *CDMA2000* standard:
  Advanced forward link trilateration (A-FLT) is standardized by 3GPP2 for CDMA2000 (3GPP2 2004). Unlike GSM or UMTS, *CDMA (IS-95)* and CDMA 2000 are time-synchronized systems. Therefore, time-difference measurement is easier. The basic idea of the A-FLT method is to measure the time difference (phase delay) between CDMA pilot signal pairs. Each pair consists of the serving cell pilot and a neighboring pilot. The time difference is converted to the range information. Finally, the range data is used to form certain curves at which an intersection is defined for the MT location.
  Although the name of this method implies that A-FLT is a MT based solution, the location can be determined either at the MT or at the network. For an MT based solution, the MT must determine the time difference of arrival among multiple pilot signals through its searcher. For an MT-assisted solution, the pilot signal measurement message along with the round-trip delay can be used to determine the time differences. Since the basic principle of this method is TDOA, the navigation equation can be solved as shown for EOTD or OTDOA.
  In addition to A-FLT, GNSS, AGNSS, and hybrid A-FLT and AGNSS position determination has been specified by 3GPP2 (3GPP2 2004).
- The Open Mobile Alliance (OMA 2013) has standardized the *SUPL* protocol (OMA 2011). SUPL enables the transfer of assistance and positioning data over an IP bearer network in the user plane for network and *SUPL enabled terminal (SET)* based positioning of a SET. The protocol defines reference points between the *SUPL location platform (SLP)* and the SET and between the *SUPL location center (SLC)* and the *SUPL positioning center (SPC)*. SUPL also defines functions for security (e.g., authentication, authorization), charging, roaming, and privacy. It utilizes existing standards where available and possible. Further, SUPL is designed to be extensible to support future positioning technologies as needed.
  Currently the following SUPL bearer networks are supported (OMA 2011): GSM and its enhancements GPRS and EDGE, UMTS, LTE, CDMA, CDMA2000 and its enhancement HRPD, WiMAX, WLAN, *I-WLAN* (3GPP/3GPP2WLAN interworking), *I-WiMAX* (3GPP/3GPP2WiMAX Interworking), and fixed broadband (e.g., cable, *DSL*, etc.).
  Several positioning methods of these SUPL bearer networks are supported by SUPL (OMA 2011): Autonomous (stand-alone) GPS, assisted GPS, EOTD (GSM), OTDOA (UMTS), A-FLT (CDMA/CDMA2000), E-CID, autonomous *Galileo and additional navigation satellite systems (GANSS)*, assisted GANSS, OTDOA (LTE), SET based E-CID, SET based OTDOA, high accuracy AGNSS, SET assisted and based short range nodes, or other sensors.
- GNSSs are wireless positioning systems. Smartphones nowadays all implement autonomous GPS and assisted GPS receivers and in the near future autonomous

GNSS and AGNSS receivers. Thus, the open service GNSS signals, which these receivers use, are standardized and specified in public *interface control documents*, see Chapter 1 for more details.

- New technologies transmitting in the 60 GHz ISM band have been recently standardized and developed (ECMA 2010; IEEE 802.11ad 2012; IEEE 802.15.3c 2009; WirelessHD 2010). These standards can be considered UWB systems as their transmission bandwidths are larger than 500 MHz (FCC 2002; ITU-R 2006). Although none of the standards considers localization, they all consider beamforming to extend the communication range. Thus, new devices can be developed that measure AOA or AOAD for localization.

9

Applications

In this chapter, we introduce applications that use the location information obtained from wireless position information. First, in Sections 9.1, 9.2, and 9.3 we consider applications that improve wireless communications systems in terms of macro diversity, radio resource management, and mobility management. A lot more information on these applications can be found at (WHE n.d.). A second major application, which has particularly driven the standardization of wireless positioning in cellular mobile communications systems is emergency calls (see Section 9.4). Finally, we give a brief overview on the wide field of LBS in Section 9.5.

9.1 Macro Diversity

Macro diversity is a spatial diversity scheme with multiple antennas being separated by several orders of wavelengths. In cellular networks or WiFi, macro diversity implies that the antennas are located at different BSs or APs. Examples of macro diversity schemes include the CDMA softhandover in UMTS (ETSI 2010) or cooperative diversity in 3GPP LTE-Advanced (3GPP 2010). In Section 9.1.1, we discuss how to apply spatial diversity schemes and space-time codes in order to provide cellular diversity that outperforms pure macro diversity with the help of location information. Next, we consider location-based synchronization for cellular OFDM in Section 9.1.2, that is, multi-link synchronization, which is a necessary prerequisite for macro diversity schemes. Finally, we investigate position aware adaptive communications systems that use adaptive modulation and coding in combination with location information to achieve macro diversity.

9.1.1 Cellular Diversity

For LTE-Advanced, significant improvements in the *SINR* at the mobile terminal are required (WHERE D3.1 2009). The concept of *coordinated multi-point (CoMP)* transmission or reception is envisioned, which implies a coordination of the transmission from multiple transmission points in the downlink (3GPP 2010; Parkvall et al. 2008). In the sequel a method is presented to take advantage of the constellation of neighboring BSs serving the same area, namely their

Positioning in Wireless Communications Systems, First Edition. Stephan Sand, Armin Dammann and Christian Mensing.
© 2014 John Wiley & Sons, Ltd. Published 2014 by John Wiley & Sons, Ltd.

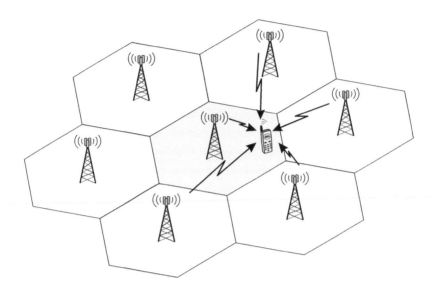

Figure 9.1   MT at cell edge under influenence from neighboring BSs

cell borders and the knowledge of the user terminal positions, see Figure 9.1. The methods apply transmit diversity techniques within these broadcasted regions. This results in a further source of diversity in addition to the existing macro diversity in such broadcasted areas. There-fore, macro diversity and transmit diversity techniques are combined that transform to cellular diversity (Plass 2008). Two transmit diversity techniques can be applied by using adjacent BSs: *Cyclic delay diversity (CDD*, Dammann and Kaiser 2001) resulting in *cellular CDD (C-CDD)* and space time block coding, that is, Alamouti coding (Alamouti 1998), resulting in the *cellular Alamouti technique (CAT)*.

9.1.1.1   Cellular Cyclic Delay Diversity–C-CDD

Transmitting the same signal from several base stations including cyclic delays will be observed as a channel with higher frequency selectivity at the receiver. For instance, this resulting additional frequency diversity can be collected by a channel code. There exists no rate loss for higher number of transmit antennas or BSs. Furthermore, there are no requirements regarding constant channel properties over several subcarriers or symbols and transmit antenna, or BS numbers. The basic principle of C-CDD is shown in Figure 9.2.

9.1.1.2   Cellular Alamouti Technique–CAT

The concept of transmit diversity using the *space-time block codes (STBCs)* from orthogonal designs, namely the Alamouti code (Alamouti 1998) can be also applied to a cellular envi-ronment. Since the space dimension in STBCs is given by multiple transmit antennas at the transmitter, the space dimension can be also shifted to simultaneously transmitting BSs in a cellular communications system. The use of the simple Alamouti code as a special case of

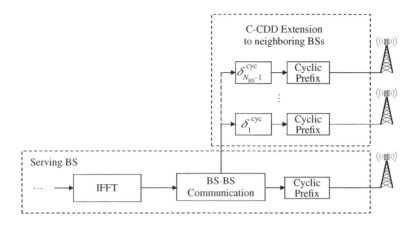

Figure 9.2 Basic principle and block diagram of C-CDD

STBCs leads to the CAT. Two mobile users can be very efficiently served simultaneously. Figure 9.3 shows two mobile users situated within the broadcast region. Both users' data are transmitted with the following signal design based on Alamouti coding:

$$B_{CAT,2} = \begin{pmatrix} S_0^{(0)} & S_0^{(1)} \\ -S_0^{(1)*} & S_0^{(0)*} \end{pmatrix},$$

where $(\cdot)^*$ represent complex conjugate, $S_0^{(0)}$ the transmitted symbol at time index 0 of the first user, and $S_0^{(1)}$ represents the transmitted signal for the second user at time index 0.

For STBCs from orthogonal designs, channel estimation for each transmit antenna at the receiver is mandatory. Disjointed pilot symbol sets for the transmit antenna branches can guarantee a separate channel estimation for each BS (Inoue et al. 2002). Therefore, the receiver has to apply a new channel estimation process for exploiting the additional transmit diversity. Furthermore, the additional pilot signals of the second transmitted signal of the neighboring BS reduce the overall throughput.

9.1.1.3 Exploitation of the Positioning Information for Cellular Diversity

Users with similar demands in the broadcasted region can benefit from the two techniques. Inter-BS communication is necessary to guarantee the transmission of the desired signals. Therefore, information of the user position would be very beneficial to initialize the cellular diversity techniques at neighboring BSs beforehand. Furthermore, with positioning information, power adaptive transmission schemes can be applied to broaden the diversity area at the critical cell border.

Two Simultaneously Served Mobile Users with CAT
In the case of CAT, two mobile users can be very efficiently served simultaneously as shown in Figure 9.3. The base stations exploit information from a feedback link that the two users are in a similar location in the cellular network. By this, both users are served simultaneously avoiding any inter-cell interference between each other and exploiting the additional diversity gain.

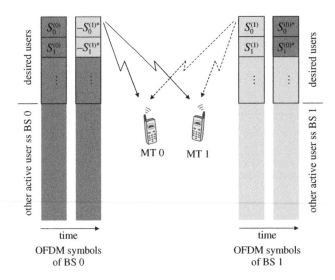

Figure 9.3   OFDM symbol design for CAT serving two users simultaneously

Positioning-Dependent Transmit Power Adaptation for Cellular Diversity

Due to the influence of the propagation path loss, the maximum diversity received from all involved BSs for cellular diversity techniques is achieved directly at the cell border. Therefore, the idea is to broaden the maximum diversity area by adapting the transmit powers regarding the position of the mobile user. This can be achieved by a transmit power adaptation to guarantee a maximum received signal strength from all BSs at the mobile.

Since the BSs are limited to a maximum transmit power P_{max}, the adaptation of the transmit power $P_{t,m}$ of each BS m in cellular diversity techniques has to follow the constraint

$$\sum_{m=0}^{N_{BS}-1} P_{t,m} = P_{max}.$$

The maximum transmit power is assumed to be normalized to one and an adaptive power factor $a_m \in [0, 1]$ regulates the transmit powers of each BS,

$$\sum_{m=0}^{N_{BS}-1} E\left\{ |a_m|^2 \right\} \overset{!}{=} 1.$$

In the following, we explain a basic power adaptation for two involved cells. The combined signals from the two base stations should be received with a maximum power of P_{max}. With the assumptions in the previous two equations, the adaptive power weighting factors are chosen by

$$a_0 = 1 - a_1.$$

The decision for using a power adaptation or just transmit the desired signal with full maximum power from only one base station is given by the intersection of the performance of reference system applying no cellular diversity technique and the performance of the system with cellular diversity and halved transmit powers, see the following evaluation section. This decision can be also given by a direct positioning information.

9.1.1.4 Methodology, Assumptions, and Assessment Criteria

The cellular network may provide input regarding the position accuracy of a mobile user in a certain location. This can be used to initiate the cellular diversity technique beforehand. The more precise the position information the better the adaptation will be. On the other side the information about the existing environment is also necessary. For instance, could one of the cellular diversity signals be totaly blocked? Therefore, is it meaningful to start the initialization of the cellular diversity techniques? Furthermore, a regular update of the position information is necessary to apply an efficient power adaptation of the transmitted signals.

The cellular diversity techniques are applied in a multiple cellular environment with two involved cells. The performance results are given for one desired user in terms of its BERs.

To evaluate cellular diversity techniques with position information we consider perfectly updated position information. Therefore, transmit power adaptation can be applied on top of the cellular diversity to broaden the received diversity area.

9.1.1.5 Evaluation

Simulation Setup

The simulations are based on the downlink and the system and environment parameters in Tables 9.1 and 9.2.

Difference between C-CDD and CAT simulation setup. Instead of a fully inner interleaving of the OFDM symbols on one OFDM frame (two-dimensional interleaver) in C-CDD, for CAT the simulation chain includes only a frequency interleaver (one-dimensional interleaver) to maintain the time non-selective constraints of the Alamouti coding. Since a mobile radio channel is more likely to be time non-selective, that is, the fading is almost constant during

Table 9.1 System parameters

Carrier frequency	5 GHz
Signal bandwidth	82 MHz
Subcarrier spacing	49 kHz
FFT length	2048
Used subcarriers	1664
One user occupies	208 subcarriers
OFDM symbols per frame	16
Guard interval length	128
Convolutional code	rate $1/2$; $(561, 753)_8$
Modulation	4-QAM
SNR	5 dB
Channel state information	perfect at the receiver
C-CDD cyclic delay	30 samples

Table 9.2 Environment parameters

Cell radius	300 m
Number of cells	2
Path loss decay factor	3.5
Channel model	IEEE 802.11n (IEE 2004), representing large open space (indoor and outdoor) statistically independent channels with equal stochastic properties from each BS
Transmit power	normalized

two successive symbols. With a two-dimensional interleaver the quasi-static assumption of the fading would be violated. Therefore, the beneficial Alamouti code would be destroyed.

Results

In the following, we present simulation results for the cellular diversity techniques applied in a two-cell scenario. Figure 9.4 represents the *bit error rate (BER)* versus the distance of the mobile user to the desired BS in meters. The cell border is at 300 m.

The performances of the applied cellular diversity methods are compared with the *orthogonal frequency division multiple access (OFDMA)* reference system using no transmit diversity technique and with a random independently chosen subcarrier allocation in each cell site. The reference system is half and fully loaded. The cellular diversity techniques transmit with halved

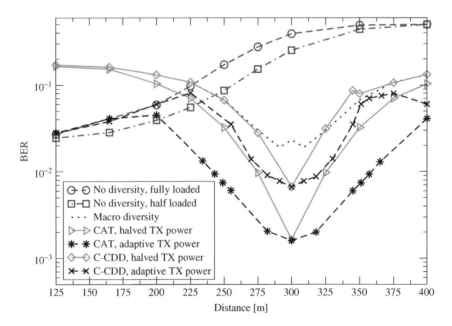

Figure 9.4 BER versus distance of mobile to the desired BS in meters for a SNR = 5 dB using cellular diversity techniques (C-CDD and CAT) with halved and adaptive transmit power and no transmit diversity technique. The cell border is at 300 m

transmit power from each BS. Thus, the total designated received power at the mobile terminal is equal to the conventional OFDMA system. Furthermore, the adaptive transmit power strategy is investigated in the following simulation results.

We observe a large performance gain in the close-by area of the cell border ($d = 300$ m) for the new proposed cellular diversity techniques. The pure macro diversity scenario at the cell border also represents the conventional OFDMA single-user case without any inter-cellular interference. C-CDD enables an additional performance gain compared to pure macro diversity, which transmits the identical signals from both cell sites without any cyclic delay. There exists a drawback of C-CDD due to the interaction between the artificial included delay and the inherent geographical delay. These two delays cancel each other out and transmit diversity does not exist anymore, which can be seen at $d = 345$ m. This drawback of the basic C-CDD principle can be eliminated by an approach using randomized delays out of a defined interval. This needs no additional signaling to the receiver. The random choice of the cyclic delay is included in the adaptive transmit power approach. Each BS is restricted to a maximum transmit power in a regulated cellular system. The received diversity and the resulting performance gain are increased in a wider area around the cell border. In contrast to the non-adaptive C-CDD with halved transmit powers at both BSs, the power adaptation guarantees the reception of the signal with maximum power throughout the whole cell.

For CAT, a larger performance gain is achieved in the cell border area compared to C-CDD. CAT exploits more additional transmit diversity at the mobile terminal on top of the already existing macro diversity. The transmit power adaptation (depending on the position of the mobile user) broadens the received diversity area as in the case of C-CDD. The decision for using the adaptive power transmission is given at $d = 225$ m for C-CDD and $d = 200$ m for CAT.

9.1.2   Location-Based Synchronization for Cellular OFDM

Wireless communications systems usually target at a frequency reuse factor of one in order to use the available spectrum efficiently. Adjacent cells access the same time-frequency resources, in particular, the same spectrum. At the cell border, downlink signals from adjacent base stations are received with similar power levels. This results in severe interference and makes it extremely difficult to achieve high data rates at cell edges if there is no coordination between the serving and interfering base stations at all.

To overcome this problem, concepts like interference coordination (see Astely et al. 2009 or Section 9.2.1), which reduce interference by coordinated resource management at the cell border, or macro diversity, where mobile terminals at the cell edge are served from multiple base stations (Plass 2008), have been proposed (see Section 9.1.1).

At the receiver side, one of the first and most important steps that must be performed is synchronization. In this process, local oscillator frequency offset as well as frame- and symbol timings have to be estimated in order to detect data symbols correctly. For synchronization purposes, a common approach is to transmit signals, which are known to the receiver. For instance, in 3GPP-LTE, such signals are transmitted within the so-called primary and secondary synchronization signals (PSS, SSS) (Tsai et al. 2007). Just like data symbols, synchronization signals are affected by interference. Therefore, it is obvious to raise the question as to whether we can use similar approaches to treat the problem of interference.

In this section, we investigate potential time synchronization performance gains resulting from the exploitation of location and channel state information. The evaluation is based on the 3GPP-LTE standard as described in Section 8.3. We have introduced in Section 8.3 the relevant aspects of the 3GPP-LTE specification for synchronization investigations and in Section 3.2.2 appropriate synchronization methods. Here, we extend them for multi-link synchronization. The proposed methods are related to the concept of macro diversity for data modulation since they exploit signal energy from different transmission sites. Besides that, many macro diversity schemes require signal synchronization from different transmission sites (see Section 9.1.1). Simulations using a typical urban multi-cell scenario based on the WINNER channel model (see Section 8.3) show the achievable performance of the proposed time synchronization methods with focus on operation at the critical cell edge.

9.1.2.1 Multi-Link Synchronization–Usage of Location Information

As mentioned at the beginning, there are high interference levels at the cell edges of a cellular communications system with a frequency reuse of one. Signals coming from different BSs superimpose at the cell border with similar power levels. It is straightforward to raise the question as to whether we can use signal energy of synchronization signals received from different BSs. Figure 9.5 points out the situation and the proposed approach for two BSs, which is very much related to the principle of macro diversity.

Signals s_1 and s_2 are transmitted synchronously from a serving BS$_1$ and an adjacent BS$_2$. Because of different signal propagation delays δ_1 and δ_2 they arrive at the MT with a time shift $\Delta T = \delta_2 - \delta_1$. If nothing is known about the signal s_2, this signal hits the signal of the serving BS as interference. However, synchronization sequences are usually known. Dependent on the distances d_1 and d_2 the time shift

$$\Delta T = \delta_2 - \delta_1 = \frac{d_2 - d_1}{c}$$

can be calculated. Consequently, ΔT is also the time shift of the correlation peaks obtained from two correlators that are matched to s_1 and s_2. Knowing the time shift ΔT the correlator output signals can be combined. Generally, we assume a bank of correlators

$$CC_i(n) = \sum_{k=0}^{N-1} s_i(k) r^*(k+n), \qquad i = 1, \ldots, N_{\text{BS}}, \tag{9.1}$$

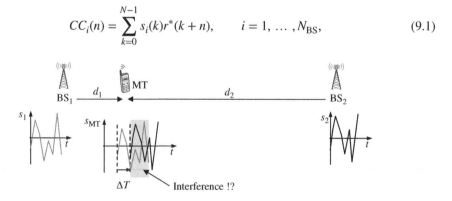

Figure 9.5   Location aware synchronization

which are matched to the synchronization sequences within radio frames $s_i(k)$, $i = 1, \ldots, N_{BS}$ transmitted from the N_{BS} BSs. The received signal in (9.1) is given by

$$r(k) = \sum_{i=1}^{N_{BS}} \alpha_i(k, d_i) s_i(k) * h_i(k - \delta_i) + n(k),$$

which is a superposition of N_{BS} signals, transmitted from adjacent BSs. Here, $s_i(k)$ denotes the signal, transmitted from BS i, that is, the 3GPP-LTE radio frames, containing the PSS and SSS sequences. $h_i(k - \delta_i)$ is the channel impulse response from BS i to the MT, taking into account the absolute signal traveling time i (in samples). "*" denotes the convolution operation. We assume that the average energy of the channel impulse (small scale fading) $h_i(k)$ is normalized to one, that is, $\sum_k \mathrm{E}\left\{|h_i(k)|^2\right\} = 1$. Large scale fading effects like shadow fading and path loss are described by the factor $\alpha_i(k, d_i)$, which is time variant and dependent on the distance d_i between BS i and the MT. We assume that the distances $d_i = cT_S\delta_i$ are proportional to the signal traveling times δ_i in samples and c is the speed of light. $n(k)$ denotes the AWGN with variance $\sigma^2 = \mathrm{E}\left\{|n(k)|^2\right\}$.

For combining the correlator signals from (9.1), we first have to shift them according to the time shifts

$$\delta_i = \frac{d_i}{cT_S}, \qquad i = 1, \ldots, N_{BS}$$

denoted in samples. The MT-BS$_i$ distances d_i are calculated from the positions of MT and BS$_i$. The positions of the BSs are fixed and usually known. The MT position can be obtained by satellite navigation systems like GPS or by positioning within 3GPP-LTE itself. Simply summing up the magnitudes of the shifted correlators results in the principle of *equal gain combining (EGC)*. In order to take into account the quality of the signal, we weight the individual correlator outputs before summation. This approach is related to *maximum ratio combining (MRC)*. Thus, we have

$$CC(n) = \sum_{i=1}^{N_{BS}} \gamma_i |CC_i(n - \delta_i)|.$$

We obtain a timing estimate by searching for the maximum of that combination, that is,

$$\hat{n}_0 = \arg \max_n CC(n).$$

The weighting coefficients are

$$\gamma_i = \begin{cases} 1, & \text{EGC}, \\ \text{SINR}_i, & \text{MRC} \end{cases},$$

where we use the SINR$_i$ as weighting coefficients. In our approach, the SNIRs are calculated from the long term fading factors $\alpha_i(k, d_i)$ by

$$\text{SINR}_i = \frac{|\alpha_i(k, d_i)|^2}{\sigma^2 + \sum_{j \neq i} |\alpha_j(k, d_j)|^2} = \frac{|\alpha_i(k, d_i)|^2}{\mathrm{E}\left\{|r(k)|^2\right\} - |\alpha_i(k, d_i)|^2}.$$

Here, we assume that $|\alpha_i(k, d_i)|^2$ represents the received signal power from BS_i without noise, that is, $E\{|r(k)|^2\} = \sum_i |\alpha_i(k, d_i)|^2 + \sigma^2$. Estimates of $E\{|r(k)|^2\}$, $|\alpha_i(k, d_i)|^2$ and σ^2 can either be provided by an appropriate channel estimation algorithm or predicted by using a fingerprint database and the knowledge of the MT position, that is, the distances d_i.

Channel State Information

Wireless communications systems have to cope with significant multipath propagation. Thus, replicas of the transmitted signal arrive at different time instances at the receiver. This results in a bias of the maximum magnitude of the cross-correlation functions as defined for instance in (9.1). To get rid of this bias, we have to cross correlate the received signal $r(k)$ with the hypothetic signal right after multipath propagation. Therefore, we replace $s_i(k)$ in (9.1) by $\tilde{s}_i(k) = s_i(k) * h_i(k)$. For multi-link synchronization, the channel impulse response $h_i(k)$ has to be provided by a channel estimation algorithm for each BS.

9.1.2.2   Results

For the evaluation of the benefit of location and channel state information, we use a cellular simulation environment setup as shown in Figure 8.13 in Section 8.3.3 and described there.

Multi-Link Synchronization

Figure 9.6 compares the synchronization accuracy at the critical cell edge, taking into account location information according to (9.1.2.1). For comparison we have drawn the graph for the single-link synchronization performance of CC. There is a marginal improvement for MRC taking into account 2 BSs. The long term fading coefficients $|\alpha_i(k, d_i)|^2$, which are required for the calculation of the MRC weighting coefficients as well as the MT position are assumed to be perfectly known.

In comparison to the previous results, Figure 9.7 shows the performance improvements taking into account additionally the channel impulse response. Compared to the single-cell synchronization performance, now multi-link synchronization provides a clear performance improvement for MRC. As it can be expected, the performance of MRC is monotonically increasing with an increasing number of combined BSs. Comparing the results, shown in Figure 9.6 and Figure 9.7, indicates that the multipath correlation bias significantly conceals potential gains resulting from the combination of synchronization signals transmitted from different BSs.

Effect of Imperfect Location Information

Previously, we have assumed that the position of the MT is perfectly known. Location information is required to calculate the timing offsets for correlation signal combination in (9.1.2.1). However, there are inaccuracies, since the MT position has to be estimated in some appropriate way. In the following, we are interested in the influence of location estimation errors on the performance of the location aware synchronization algorithms. Compared to the previous results, we have modeled and introduced a location error as a two-dimensional Gaussian random process with zero mean and standard deviations of 10 and 100 m. Results are drawn for

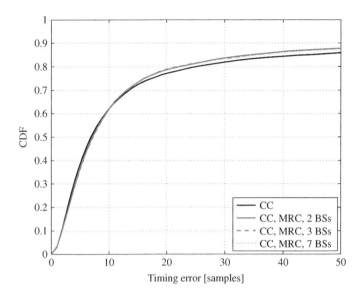

Figure 9.6 Multi-link SSS synchronization accuracy at the cell border for *CC* with MRC for the 2, 3, or 7 strongest BSs. Note that different MRC curves coincide

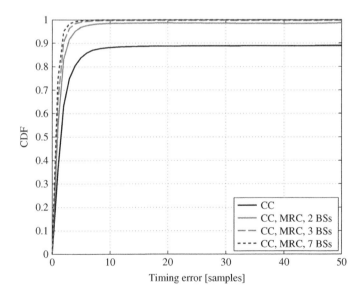

Figure 9.7 Multi-link SSS synchronization accuracy at the cell border for CC with MRC for the 2, 3, or 7 strongest BSs, taking into account the channel impulse response

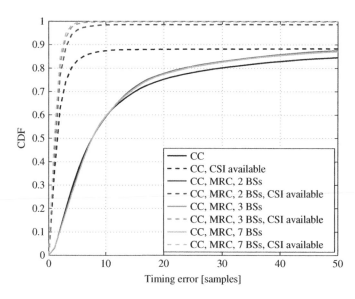

Figure 9.8 The effect of location inaccuracy (= 10 m) on SSS synchronization performance at the cell border for CC with MRC for the 2, 3, or 7 strongest BSs

MRC with and without small scale *channel state information (CSI)* in the form of the channel impulse response. For a location error with $\sigma = 10$ m depicted in Figure 9.8 for MRC, simulation results show a marginal degradation compared to perfect location knowledge both with and without availability of CSI. For a location error with $\sigma = 100$ m shown in Figure 9.9, performance degradations can be observed for MRC taking into account CSI. Nevertheless, the gain in comparison to the single-cell case is still observable in the availability. This means, the multi-link CDF curves earlier approach a probability of 1. For MRC without CSI, the multipath bias effects seem to be dominant. So, the degradation due to location inaccuracy is mainly affecting the estimation accuracy, which increases the 90% timing error from 3–4 samples to 12–14 samples. However, it has no significant influence on the availability.

To conclude this section, we have investigated the benefit of location and channel state information for synchronization in a cellular wireless communications system. The basic methods used for time synchronization have been introduced in Section 3.2.2 and extended for multi-link processing in this section, where synchronization signals transmitted from several adjacent base stations can be exploited in order to increase synchronization performance. This is a prerequisite for many macro diversity schemes (see Section 9.1.1). The availability of location information allows for timely relating synchronization signals received from different base stations and, accordingly, their corresponding correlator signals at the receiver. Simulation results have shown significant timing performance gains if small scale fading information, that is, the channel impulse response, is available and taken into account for correlations. Reversely, this means that timing degradations due to multipath propagation significantly conceal potential gains resulting from location aware multi-link synchronization. Therefore, the combination of location information and channel state information promises remarkable performance gains for high precision time synchronization requirements exploiting macro diversity.

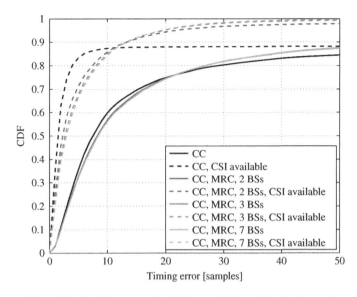

Figure 9.9   The effect of location inaccuracy (= 100 m) on SSS synchronization performance at the cell border for CC with MRC for the 2, 3, or 7 strongest BSs

9.1.3   Position Aware Adaptive Communications Systems

In recent years, operators have experienced strong growth in data services (Cisco 2012). Due to the rapidly growing mobile data communications, future cellular communications systems need high spectral efficiency over limited bandwidth to provide high data rates under the constraints of high spectrum costs and spectrum scarcity. *4G* systems, such as *3GPP-LTE* (ETS 2009), achieve high spectral efficiency through *adaptive modulation and coding (AMC)* exploiting *CSI at the transmitter (CSIT)* in combination with multiple-input and multiple-output, macro diversity, and cooperative communication techniques. These techniques exploit the spatial dimension of the wireless channel to increase spectral efficiency. Therefore, the positions of BSs, relays and MTs matter. To obtain the positions of these communication devices, current positioning systems are used that include but are not limited to GNSSs, cellular communications systems, and WiFi systems (see Chapters 1 and 8).

A main disadvantage of AMC exploiting CSIT in 4G systems is that AMC suffers severely from outdated CSIT. Hence, we present in this subsection position aware adaptive cellular communications systems, which can overcome this drawback. The main focus of this subsection is a performance assessment of position aware adaptive cellular communications systems exploiting macro diversity compared to state-of-the-art AMC systems exploiting macro diversity that use only feedback or predict CSIT in addition to the feedback CSI. Both systems exploit CSIT under ideal conditions. By ideal conditions we mean that there are no errors in the CSI feedback or information loss except for a feedback delay, that is, the feedback CSI can be outdated. Furthermore, for position aware systems we assume perfect position knowledge and a fingerprint database that is perfectly up-to-date. Numerical results demonstrate that position aware adaptive systems operate close to the channel capacity over a large range of feedback delays and channel variability when exploiting macro diversity. Furthermore, position aware adaptive communications systems achieve large capacity gains compared to state-of-the-art AMC systems exploiting macro diversity for medium to large feedback delays.

9.1.3.1   Position Aware versus Position Unaware Adaptive Communications Systems

Before introducing the systems framework let us consider an example to highlight similarities and differences between adaptive communications and position aware adaptive communications systems. Figure 9.10 shows two BSs that employ adaptive beamforming to serve three MTs.

In a classic adaptive beamforming scheme, the MT periodically reports its CSI conditions to the BSs. The BSs then adjust the beamforming weights based on past and current CSIT (Godara 1997a, b). By adjusting the beams, the BSs implicitly track the position of the MTs as the CSI of the MTs is position dependent.

On the other hand, if the MTs know their position and report it periodically back to the BSs, the BSs need a mapping between position and CSIT. This could be given by some propagation equations such as the simple $1/r^2$ law for free space propagation or via CSI measurements stored in a fingerprint database (see Section 2.3). For a fingerprint database, the position is then equal to the database search key. Thus, the BSs can adjust their beams according to the position and a mapping of the position to the CSIT.

However, if we compare these two approaches, they are fundamentally the same. So what could be the benefit of exploiting position information in a communications system?

9.1.3.2   Proactive Information in Position Aware Adaptive Communications Systems

To our understanding, a communications system can benefit from position information if it is proactive. Thus, a position aware communications system needs to exploit not only short-term channel coherence, but also mid-term or long-term coherence of the user's position and movement.

For instance, let us consider the example in Figure 9.11. If we assume an MT that regularly reports its position to the BSs, we can see that the position uncertainty in Figure 9.11(a) and with that the CSIT uncertainty increase over time (see Figure 9.11(b)). When the user moves into the crossing and reports its position as in Figure 9.11(c), the BSs would predict from the past that there exist three possible choices for the user to proceed. This implies a high uncertainty in the future position and thus, in the future CSIT. Hence, it will be very difficult for

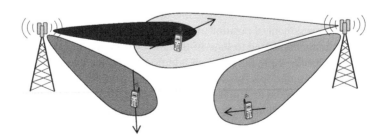

Figure 9.10   Adaptive beamforming with and without position information

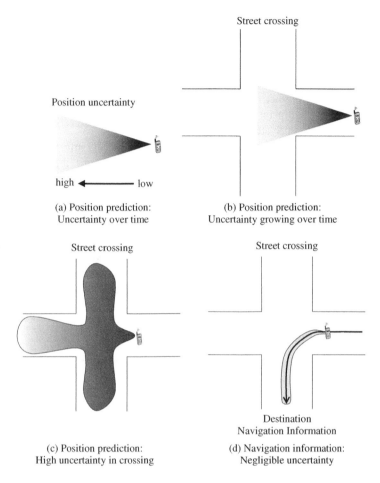

Figure 9.11   Proactive information: Position prediction and navigation information

the BSs to adapt to the correct future CSIT, which may change abruptly due to the user walking around a street corner. On the other hand, the MT cannot only report its current position back to the BSs, but also its navigation route and destination (see Figure 9.11d). With this information in combination with predictions on the user velocity, the BSs predict that the user will move on the displayed trajectory. Hence, the BSs are able to adapt to the correct future CSIT. Thus, the BSs proactively exploits navigation information, that is, the navigation destination and route.

9.1.3.3   Systems Framework

In the following we present a systems framework that distinguishes between four systems:

- *System I* is non-adaptive and cannot exploit any channel coherence. It only uses present CSI, for example, in the receiver for coherent demodulation. An example of such a system is a *fixed modulation and coding (FMC)* scheme.

- *System II* is adaptive and can exploit channel coherence. It uses present and past CSI and can short-term predict future CSIT limited by the channel coherence. Two examples of such a system are *AMC with CSI feedback* and *AMC with CSI feedback and CSIT prediction*.
- *System III* exploits past and present positioning and mobility information. It is adaptive and exploits channel and spatial mobility coherence. Furthermore, it uses present and past CSI and can mid-term predict future CSIT by exploiting a mapping between position and CSI, for example, a fingerprint database. *AMC with position prediction and a fingerprint database* is an example of System III.
- *System IV* exploits proactive navigation, positioning, and mobility information. It is adaptive and exploits channel and spatial mobility coherence. Furthermore, it uses future, present, and past CSI, and can long-term predict future CSIT. For example, *AMC with navigation information and a fingerprint database* is a realization of System IV.

Among other reasons, we introduce this system framework to benchmark the position unaware Systems I and II against the position aware Systems III and IV. The main difference between Systems I and II and Systems III and IV is the mid-term to long-term future CSIT prediction. Through this prediction Systems III and IV use the larger coherence of the position, mobility, and navigation information to reduce uncertainty in the prediction.

9.1.3.4 Capacity Evaluation

In order to asses the performance of the different systems presented in this section numerically, we conducted the following capacity evaluation:

1. We determine the mean SNR for the channel between each BS and MT at position (x,y) and select the BS-MT link with the largest SNR.
2. We check if the channel between BS and MT at position (x,y) is LOS.
3. If the channel is *LOS*, we approximate the channel capacity by the *AWGN channel capacity* for the mean SNR $\bar{\gamma}$, that is,

$$C\left(\bar{\gamma}\right) = \mathrm{ld}\left(1 + \bar{\gamma}\right).$$

4. If the channel is *NLOS*, we approximate the channel capacity by the *Rayleigh channel capacity* as

$$C\left(\bar{\gamma}\right) = \mathrm{E_h}\left\{C_{AWGN}\left(|h|^2\bar{\gamma}\right)\right\},$$

   where h denotes the CSI at the MT position (x,y).
5. Given a system as in Section 9.1.3.3, we check if the transmission rate R at the MT position (x,y) is below or equal to the capacity $C\left(\bar{\gamma}\right)$:
   - If $R \leq C\left(\bar{\gamma}\right)$, we have a successful transmission.
   - If $R > C\left(\bar{\gamma}\right)$, we have a transmission outage.

This procedure is then repeated and averaged over several user tracks as shown in Figure 9.12. Note that the first step is providing macro diversity as it is using the BS-MT link with the best SNR similar to antenna selection diversity (Paulraj et al. 2003).

In the following, we analyze this procedure for the four different systems. As *System I* is non-adaptive, it uses an FMC scheme. The transmission rate of *FMC* is $R = m \cdot R_C$, where m

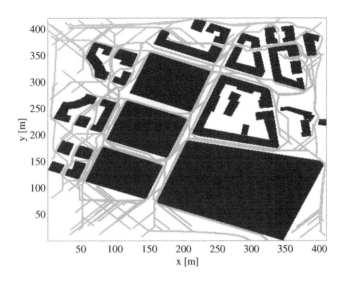

Figure 9.12   Track realizations in an urban environment

denotes the number of bits per symbol for a modulation alphabet and R_C the code rate. If the transmission rate R of FMC is smaller than or equal to the channel capacity $C\left(\bar{\gamma}\right)$, an error free transmission is possible. Otherwise, the transmitted bits cannot be decoded error free. In this case, we discard the transmitted bits. Here, we choose the transmission rate of FMC to maximize the mean capacity given the complete CSI of the scenario.

System II employs *AMC*. In this case, the MT feeds back CSI to the BS. The performance limiting factors of AMC are the feedback delay and the amount and quality of feedback CSI required by the adaptive transmitter algorithms. Examples for feedback CSI include estimated channel coefficients, SNR, and propagation conditions such as LOS or NLOS. According to the feedback CSI, the BS selects the modulation type and code rate. Here, the initial values for AMC and FMC are the same. As in the case of FMC, a communication outage occurs if the current transmission rate R according to the last reported CSI is larger than the channel capacity $C(\bar{\gamma})$. Otherwise, an error free transmission is possible. Clearly, the achievable capacity for AMC will degrade with increasing feedback delay as the CSIT becomes faster outdated.

Further, we can define a *System II* that *predicts CSIT* in addition to AMC and CSI feedback. Here, the critical parameter is the temporal channel coherence. The stronger the channel is correlated over time, the more accurate the CSIT prediction will be. Typically, the channel coherence time is defined as the time during which the channel is correlated above a threshold ρ_{coh} (Fazel and Kaiser 2003). Within the channel coherence time, the BS can predict the CSIT very accurately. Hence, the BS can adapt the transmission rate R to the channel capacity $C(\bar{\gamma})$. If the channel coherence time becomes too small, for example, smaller than the feedback delay, the BS can switch off the CSIT prediction and can use AMC with CSI feedback.

The *System III* is the first position aware adaptive communications system that we consider. Based on past position measurements known to the BS, the BS predicts the *current and future position* of the user's MT. Thus, it can adapt the modulation and coding in conjunction with a location dependent fingerprint database that contains CSI. Compared to the third system, the critical parameter here is the temporal motion coherence (see Yuillel and Grzywacz 1988).

Within the motion coherence time, the BS can accurately predict the MT position. Thus, it can adapt the transmission rate R to the channel capacity $C(\bar{\gamma})$. If the motion coherence time becomes too small, the BS can switch back to AMC with CSIT prediction and CSI feedback.

The *System IV* exploits navigation information together with position prediction and a fingerprint database. One of the most commonly used cases for positioning is the point-to-point navigation. Here, the MT calculates the user a navigation route from his current position to his destination, which is an example of navigation information. The main benefit of System IV compared to System III is that we can obtain the MT position accurately not only when the motion coherence is high, but also when the user turns sharply around a street corner due to the navigation route (see Figure 9.11d).

9.1.3.5 Numerical Results

To evaluate the performance of these four systems numerically, we choose a cellular communications system as in Section 8.4. Figure 9.13 depicts the setup for 11 BSs from the cellular grid out of 13 active BSs. Each BS's cell is divided into three hexagonal sectors by three sectorized antennas, each one covering approximately an angle of $120°$. The inter-site distance is 1500 m. Within the cellular network, the simulation environment is located at the cell edge of three neighboring BSs in Figure 9.13. We created realistic models for the CIR with a ray tracing simulator in an urban environment (see Figure 9.12 and Section 6.2) that covers a 400×400 m area with a resolution of 1 m². The map of Figure 9.12 shows 100 random user track realizations. Each track is generated with a mobility model based on the gas diffusion algorithm presented in Section 6.4. The average building height in Figure 9.12 is 26 m.

To simplify our performance analysis, we assume that a user is moving at a constant speed through the scenario. The user's MT is kept at a constant height of 1.5 m relative to the street level. Further, we assume a noise power spectral density of $N_0 = -174$ dBm. Given the CIR for each point in Figure 9.12 by ray tracing and N_0, we can compute the SNRs, which range from -105–40 dB. With the SNR and CIR for each point, we then numerically compute the mean capacities of the different systems over all user tracks in Figure 9.12.

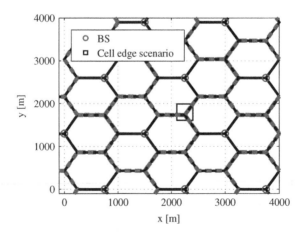

Figure 9.13   Cellular communications network

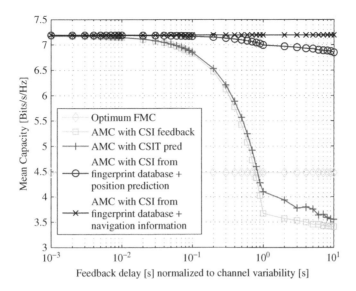

Figure 9.14   Comparison of position aware adaptive communications systems and state-of-the-art AMC systems for maximum channel or movement coherence of 0.9

Figure 9.14 presents results for the mean capacity [bits/s/Hz] versus the feedback delay [s] normalized to the channel variability [s]. We define the channel variability as

$$\Delta t_{channel} = d_{channel}/v_{user},$$

where $d_{channel}$ denotes the distance over that the channel is approximately constant and v_{user} is the user's speed.

For instance, if a practical system has a feedback delay of 10 ms, $d_{channel} = 1$ m, and $v_{user} = 10$ m/s, the normalized feedback delay is $\Delta t_{channel} = 10^{-1}$ in Figure 9.14. For this delay, the AMC is 2.35 bits/s/Hz spectrally more efficient than the optimum FMC. Note that the FMC system transmits 5.5 bits/s/Hz, which is the optimum transmission rate for the scenario depicted in Figure 9.12 resulting in a maximum mean capacity of 4.48 bits/s/Hz for FMC in Figure 9.14. The CSIT prediction only improves slightly over AMC whereas the position prediction together with a fingerprint database improves by 0.34 bits/s/Hz over AMC. Above a normalized feedback delay of 10^{-1}, AMC with and without CSIT prediction shows a severe degradation in spectral efficiency due to outdated CSIT. In contrast, the spectral efficiency of a system with position prediction and a fingerprint database degrades only slightly by 0.35 bits/s/Hz for the maximum feedback delay of 10 compared to the system that combines navigation knowledge, position prediction, and a fingerprint database, which shows no degradation at all. As the proactive position aware AMC system can predict the CSIT due to the navigation information and fingerprinting database always correctly, it coincides with the channel capacity. Hence, position aware adaptive communications systems require less frequent CSI feedback than state-of-the-art AMC systems do.

To summarize this section, we presented a systems framework for benchmarking position unaware systems against position aware systems exploiting macro diversity. The position unaware systems are System I, which is non-adaptive, and System II, which is adaptive, exploits channel coherence and predicts CSIT. The first position aware communications system is System III, an adaptive system that uses both channel and spatial movement coherence in combination with position prediction and a fingerprint database. The second position aware communications system is System IV, a proactive adaptive system that uses channel and spatial movement coherence navigation information in combination with a fingerprint database. As our numerical capacity analysis demonstrates, the position aware Systems III and IV can operate near the channel capacity over a wide range of feedback delays and channel variability.

9.2 Radio Resource Management–RRM

Besides macro diversity (see Section 9.1), CoMP can improve the SINR at the MT also through *radio resource management (RRM)*. In the following two subsections, we explain location-based *inter-cell interference coordination (ICIC)* (WHERE D3.4 2010) and a location-based *relay selection* scheme (WHERE D3.6 2010) to improve the SINR.

9.2.1 Location-Based Inter-Cell Interference Coordination–ICIC[1]

Due to the frequency reuse factor of 1 in LTE and LTE-A (ETS 2013b), cell edge MTs in particular will suffer from inter-cell interference. Therefore, ICIC schemes are mandatory to increase the cell-edge throughput. Indeed, cell-edge users or MTs suffer from a strong inter-cell interference, especially in interference-limited scenarios with low inter-site distance (WHERE D3.4 2010). Since a trend is to reduce the cell size, it becomes crucial to control the inter-cell interference in order to achieve better cell-edge throughput. Furthermore, operators would like to provide a more homogeneous service, with better balanced cell-edge and cell-center throughputs. Figure 9.15 illustrates how the inter-cell interference impacts the SINR level. MT1 communicating with BS1 is a cell-edge MT, receiving a weak useful signal S1 and a strong interference I1 from the neighboring BS2. On the other hand, MT2, also communicating with BS1, is a cell-center MT, receiving a strong useful signal S2 and a weak interference I2.

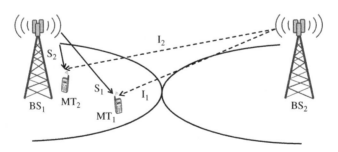

Figure 9.15 Inter-cell interference in a cellular system

[1] Source: (WHERE D3.4 2010), Chapter 5 'Location-based Inter-cell Interference Coordination'. Reproduced with permission of Loïc Brunel, Nicolas Gresset, and Mélanie Plainchaulti.

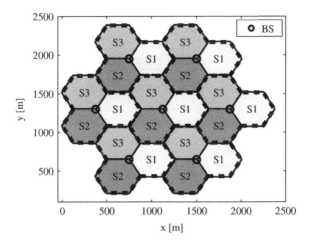

Figure 9.16 Planning with three sector cell types S1, S2, S3

Thus, MT1 experiences a low SINR whereas MT2 benefits from a high SINR. It results in strong throughput disparities among MTs. The main purpose of ICIC is to reduce these disparities. However, a major issue when applying ICIC is the difficulty anticipating the interference level. Indeed, it depends on the actual time-varying scheduling in neighboring cells.

This is why radio resource planning is needed among neighboring cells in order to coordinate inter-cell interference on a long-term basis. The radio resource planning may be fixed or may vary in time due to some signaling between BSs as in LTE. In this section, we focus on a fixed intra-band frequency planning with three sector-cell types S1, S2, and S3, as depicted in Figure 9.16 for a hexagonal deployment of seven BSs with three sectors each.

A given sector type, corresponding to a given power pattern, varying in frequency but fixed in time, is associated to each sector. The sector type S_i is chosen for a given sector i in such a way that all its neighboring sectors and cells have a sector type different from S_i. There exist different techniques of interference coordination, depending on the pre-defined power pattern and the coordination strategy. In the sequel, we study different approaches taking into account the impact of MT location knowledge on the ICIC performance. The positioning information can be directly used for ICIC or be an entry for a fingerprinting map.

9.2.1.1 Hard Frequency Reuse

With the so-called hard frequency reuse strategy, signals of neighboring sectors are transmitted in non-overlapping frequency sub-bands. With three sector types per cell, each sector is allocated a third of the system bandwidth, which by definition results in a frequency reuse factor 3 as shown in Figure 9.17. The throughput is increased by the SINR improvement resulting from the interference mitigation. Unfortunately, it is also limited by the partial bandwidth usage that induces both a direct division of the spectral efficiency by a factor 3 and a degraded performance due to larger restrictions on the frequency scheduling.

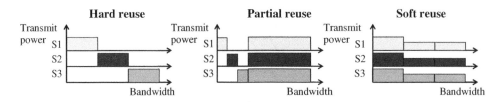

Figure 9.17 Hard, partial, and soft frequency reuse for ICIC

9.2.1.2 Partial Frequency Reuse

In order to circumvent the drawbacks of the hard reuse strategy, the so-called partial frequency reuse (Sternad et al. 2003) applies a frequency reuse factor of 3 with the three sector-cells on a part of the bandwidth and a frequency reuse factor 1 on the rest of the bandwidth (see Figure 9.17). Cell-edge MTs are scheduled in the part of the bandwidth with frequency reuse 3, where they experience a limited interference. However, they suffer from the drawbacks of hard frequency reuse. Thus, the performance of partial frequency reuse lies between the performance of hard frequency reuse and the performance without ICIC. The lower the proportion of MTs in hard reuse is, the closer to the no ICIC case the partial reuse performs.

9.2.1.3 Soft Frequency Reuse

In order to avoid the spectral efficiency division when using a frequency reuse factor 3, soft frequency reuse has been introduced (Huawei 2005; Xiang et al. 2007). In each cell, a large maximum transmission power P_{max} is allowed in a single sub-band, whereas only a limited maximum transmission power is allowed in the two other sub-bands (see Figure 9.17). We denote β the ratio P_{min}/P_{max}. Based on this long-term power frequency planning, the BS is able to perform an efficient scheduling in order to increase the cell-edge throughput, at the expense of the average throughput. Cell-edge MTs are scheduled on the sub-band with improved SINR, that is, the sub-band with transmission power P_{max} and lowest experienced interference, whereas cell-center MTs are scheduled in the two other sub-bands. The choice of β impacts performance. When β decreases from 1, the cell-edge performance is first improved. Then, β becomes so low that original cell-center MTs are so degraded that they have a spectral efficiency lower than original cell-edge MTs and the performance starts to decrease. Thus, there is an optimal β value. The scheduler must first determine if a MT is a cell-edge MT or a cell-center MT. We investigate two types of classification. In the long-term classification, MTs, which belong to the third of MTs with lowest long-term SINR, are classified as cell-edge MTs. Fast fading is not taken into account here. In short-term classification, the frequency-selective fast fading is included in the SINR evaluation. This more accurate classification further improves the cell-edge throughput.

Use of Location Information
There is a clear relationship between the long-term SINR level and the MT location, without shadowing. The SINR level depends on the neighboring BSs' positions, which should be made available. However, a shadowing with a strong variance makes the location information irrelevant (WHERE D3.1 2009). With shadowing and long-term classification, a fingerprinting

map containing the MT state, that is, cell-edge or cell-center, for a given position may be used for long-term classification. With such a map, shadowing is efficiently taken into account, especially because it is spatially correlated. The expected advantage of using the location information is low complexity. Once the fingerprinting map has been built, the scheduler directly knows if the MT is a cell-edge MT or a cell-center MT from its position. It does not have to accumulate short-term SINR values reported by the MT to compute the long-term SINR before deciding if the MT is a cell-edge MT or not. For short-term classification, a fingerprinting map containing the experienced long-term interference can be used together with the instantaneous power received by the MT from its serving BS. The latter power is obtained at the MT by channel estimation and reported to the BS. Obviously, the relevance of the fingerprinting maps depends on the location accuracy and the fingerprinting map spatial granularity. It also depends on the interference level quantization for interference maps.

9.2.1.4   Summary of Results

Using location information reduces the complexity of ICIC (WHERE D3.4 2010), limiting the required amount of reporting from the mobile terminals thanks to the usage of fingerprinting maps at the base stations. The performance results for the different techniques show that the positioning error should be lower than 50 m, that is, the shadowing correlation distance, in order to allow efficient location-based ICIC. A positioning accuracy of around 10 m brings reasonable degradation compared to ICIC with ideal location information.

9.2.2   Location-Aided Relay Selection[2]

In wireless networks, the performance of data transmissions is depending on the distance between transmitter and receiver nodes as well as on the number of collocated nodes and their activity patterns (WHERE D3.6 2010). Neighbor nodes can cause performance degradation due to interference and collision. Furthermore, when there is a large distance between the transmitter and receiver, rate adaptation schemes switch to a more robust modulation scheme, which decreases the transmission rate. As shown in (Liu et al. 2007b) this reduction in transmission rate can significantly reduce the network capacity, and thereby also transmissions with high rate may be affected. By introducing relaying techniques, nodes located between a transmitter and receiver pair can be exploited to provide a multihop path with shorter links. The shorter links between nodes may form a more reliable path to the destination node, which allows us to achieve a higher throughput (Zhu and Cao 2005).

Here, we consider the scenario where mobile users primarily need to make downlink transmissions (WHERE D3.6 2010). Audio or video streaming are examples of applications leading to such traffic patterns. In such downlink scenarios, we will therefore focus on *centralized two-hop relay selection* where transmissions from the AP may be direct or via a two-hop relay path. We rely on maintaining an up-to-date view on the potential relay nodes, but for the downlink case, only the AP needs to have an updated view.

The AP's ability to determine the best path depends on the accuracy of the AP's view on the link properties in the network. This view is updated by periodic collection of link quality or

---

[2] Source: (WHERE D3.6 2010), Section 4.1 'Impact of Mobility and Inaccurate Path Loss Model Parameters on Relay Selection'. Reproduced with permission of Jimmy Jessen Nielsen.

position information measurements, as described in more detail later. The age and availability of link measurements in relation to the movement speed of the MT is expected to impact the accuracy of the choice of path. In addition to these factors, the number of nodes is also expected to influence the path selection. In the sequel we will present how the path selection is impacted by these factors.

9.2.2.1   Scenario Description

We consider downlink transmissions from a fixed AP to MT in an IEEE 802.11 network. Data transmissions can be done directly to the destination MT or as a two-hop transmission via any intermediate relay node, as sketched in Figure 9.18. A relayed transmission is only considered if it provides a better transmission quality than a direct transmission. In this section the goodness of a transmission is determined in terms of the achieved BER.

A specific relayed transmission path is chosen if its BER from AP to MT over the relay is smaller then the BER of the direct path from AP to MT and smaller than all other possible relay transmission paths. Notice that this approach does not ensure that the chosen relay path delivers a higher throughput than the direct path, due to the store and forward behavior of the relay node. However, the lower BER increases reliability of the transmission.

For both schemes, the procedures used to collect measurements are envisioned as data link layer protocol extension. Further, since old measurements may be misleading due to mobility of the MTs, it is assumed that a parameter denoted α_{store} exists in the AP, which expires measurements when their *storage time* exceeds α_{store}.

We can further define the age of a measurement as the elapsed time since the hello broadcast of the latest measurement for that link was initiated. The age of a link measurement is a stochastic process that is influenced mainly by the hello broadcast generating process. A random jitter is added to the inter-event time for hello broadcasts, which ensures that hello transmissions from different MTs are not in sync. Further, since the MTs' movements are independent of each other, the mobility model is assumed to be ergodic.

9.2.2.2   SNR Based Relay Selection

In order for the AP to make this decision, it needs to estimate the BER of the network links. The BER is estimated from the measurements of link SNR, which are obtained from MTs and collected by the AP.

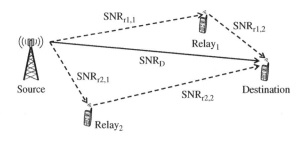

Figure 9.18   Example of possible direct and two-hop paths

By letting all MT broadcast *hello* messages periodically with average interval duration μ_{hello} [s], other devices within receiving range are able to measure the RSS and thereby the SNR of the hello broadcast. This measurement is assumed to represent the link state at this moment in time. Further, all nodes are using the same fixed transmission power. The hello message is a IEEE 802.11 MAC frame without payload (20 octets), since only the MAC address is needed for the receiver to identify the broadcast source. Notice that hello broadcasts may be lost if collisions occur.

Whenever an MT overhears a hello broadcast and thereby obtains an SNR measurement, it assembles a measurement frame and sends it to the AP using a unicast transmission. Due to the small frame size, RTS/CTS is not applied but standard 802.11 retransmissions are used if needed. The measurement frame is envisioned as being a MAC control frame that carries the MAC address of the hello broadcast source (6 octets) and the SNR measurement (2 octets), which amounts to a frame size of 28 octets when adding this information to the standard 802.11 control frame layout (IEEE 802.11 2007). Hereby, N hello broadcasts lead to $N \cdot (N - 1)$ measurement transmissions to the AP, in the case where all nodes receive all hello-broadcasts. The actual amount of measurement transmissions may vary due to losses and possible retransmissions.

The AP identifies the link from which the measurement has been obtained from the MAC addresses of the broadcast node and measurement node. Notice that it is assumed that links are symmetric.

Having obtained SNR measurements from links between devices in the network, the AP now chooses the best path.

9.2.2.3 Location-Based Relay Selection

The idea behind this scheme is that by knowing the locations of the MTs in the network, the path-loss, the SNR, and in turn, the BER can be estimated with propagation models by assuming fixed transmit power and approximating noise floor and propagation properties of the environment. Locations are obtained by letting all MTs transmit location measurements periodically with interval μ_{loc} to the AP using unicast transmissions. Similar to the case with hello broadcasts for the SNR measurement based scheme, the initial transmission time is chosen for each node uniformly random in the interval $[0, \mu_{\text{loc}}]$. Further, the following location measurement transmissions are offset with a uniform random jitter in the interval $[-0.1 \cdot \mu_{\text{loc}}, 0.1 \cdot \mu_{\text{loc}}]$ to avoid transmissions being in sync. The measurement frame is a MAC control frame that carries the longitude (4 octets) and latitude (4 octets) of the node. Assuming the longitude and latitude are given as a degree decimal fraction and the circumference of the Earth is 40 000 km, the precision that is supported by this format is approximately $\frac{40000 \, \text{km}}{2^{4\cdot8}} = 0.01$ m. The frame size amounts to 28 octets when adding longitude and latitude information to the standard 802.11 control frame layout (IEEE 802.11 2007). This is the same size as the SNR measurement frame.

Having collected the MT locations, first the path loss is estimated with this path loss model from Section 2.3.2:

$$\overline{PL}(d) \, [dB] = PL(d_0) \, [dB] + 10n \log_{10} \left(\frac{d}{d_0} \right)$$

where $\overline{PL}(d)$ is the path loss in dB at the receiver, d is the distance between transmitter and receiver, $PL(d_0)$ is the path loss in dB at a reference distance $d_0 = 1$ m, and n is the path loss exponent. The value of n is scenario dependent and its exact value is typically not known in advance. The sensitivity to inaccurate estimates of this parameter have been investigated in detail in (WHERE D3.6 2010).

Given a specific transmit power level P_{tx}, the calculated path-loss $\overline{PL}(d)$ and assumed noise floor N_{floor}, the SNR is calculated as:

$$SNR = P_{tx} + \overline{PL}(d) - N_{floor} - X [dB]$$

where X is a random variable representing shadowing due to obstacles in the environment (WHERE D3.6 2010). Having determined the SNR, the expected BER can now be calculated using theoretical expressions from reference (Proakis 2001).

9.2.2.4   Summary of Results

The scenario considered in (WHERE D3.6 2010) concerns downlink data transmissions in a IEEE 802.11 based wireless network with mobile users. The focus of this section has been on analyzing relay selection schemes when either direct or two-hop relayed transmissions are possible for each data transmissions. Specifically, an SNR measurement based scheme and a location measurement based scheme have been considered and their performances have been compared. The schemes have been evaluated for the random waypoint mobility model as well as using ns-2 and MATLAB simulations. The number of mobile devices, their speed, the measurement frequency, and lastly the measurement storage time threshold have been varied. In particular, for the location-based scheme the impact of inaccurate path loss model parameters has been investigated. Results were compared to a scheme that always uses direct transmissions and an ideal scheme that has instant and perfect link state knowledge.

The results show that for relatively fast moving mobile devices (5–15 m/s) the achieved average BER performance of the SNR measurement based scheme can get significantly worse than the always direct scheme. Increasing the hello broadcast rate can mitigate this effect, however, at the cost of a linear increase in signaling overhead. (WHERE D3.6 2010) showed that by limiting the measurement storage time, that is, letting the AP use only fresh measurements, a performance could be achieved that is slightly better than or equal to the always direct scheme without increasing the signaling overhead. This result underlines the importance of choosing the storage threshold parameter carefully in scenarios with mobile devices. However, in case of fast moving users, the measurement collection frequency that is required for acceptable performance results in a large signaling overhead.

(WHERE D3.6 2010) therefore proposed a relay path selection scheme, which uses collected location information together with a path loss model for relay selection and thereby creates considerably less signaling overhead. It has been shown that due to reduced signaling overhead, this scheme allows considerably higher movement speeds as compared to the SNR measurement based scheme. In the considered case, a four-fold increase of the movement speed was possible.

Further (WHERE D3.6 2010) found that the required location measurement accuracy was comparable to the typical accuracy of standard GPS systems. In many cases, a standard

GPS system would therefore be usable, however, as the accuracy of GPS in urban/indoor environments is typically worse, a localization systems with a higher accuracy than standard GPS, should be considered for these cases. For example, network based localization methods could be exploited to improve the localization accuracy.

As the parameters of the environment are not always known in advance, the sensitivity of the location-based relaying scheme towards inaccurate settings of parameters in the path loss model have been investigated. With regard to the path loss exponent, which is typically either unknown and thus guessed from the environment characteristics or estimated as the average over a larger area, (WHERE D3.6 2010) found that estimates within a relatively wide range of ±1 around the true value resulted in near-optimal results. In the case of a *NLOS* situation, the relaying performance was severely degraded without a priori knowledge. If on the other hand knowledge of *LOS* or NLOS between nodes was made available by extending the location-based scheme with a spatial map of obstructions, the obtained performance was useful for estimates within ±3 dB of the true attenuation factor.

9.3 Mobility Management[3]

Mobility management is a major function of every wireless, and in particular cellular, network. Its goal is to locate and track the mobile user and to enable phone calls and data traffic while the user enters and leaves cells and networks. In this section, we consider location-based handover between heterogenous networks in particular WiFi and LTE (WHERE D3.7 2010). For instance, the MT in Figure 9.19 travels through two LTE cells and a WiFi hotspot. Depending on its speed and the handover delays, the cellular and WiFi networks may execute handovers to increase the throughput that the MT experiences. The cellular network may use the location information of the MT such as current location, speed, and its movement history to predict future location information. This future location information may then be used to initiate a handover to the WiFi hotspot and the return handover to the LTE network in advance to decrease the handover delay experienced by the MT.

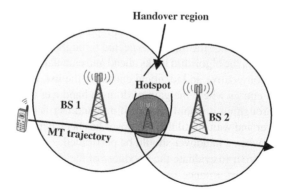

Figure 9.19 Location-based handover between heterogenous networks

[3] Source: (WHERE D3.7 2010), Chapter 3 'Location assisted hand-over prediction for WiFi and LTE'. Reproduced with permission of Jimmy Jessen Nielsen.

9.3.1 Location Assisted Handover Prediction for WiFi and LTE–Algorithm

Here, we consider location-based handover enhancements with two different communications technologies, that is, a cellular LTE network and IEEE 802.11 based WiFi hotspots (WHERE D3.7 2010). Our results will be based on these technologies; however, the proposed algorithms are general and can easily be adapted to other handover scenarios.

Assuming the throughput is higher on the WiFi network than on the cellular network, it may be beneficial to perform a handover to the WiFi network to achieve an increased throughput. However, it may not be beneficial to handover to a WiFi network if the connection is only available for a very short period of time. In this case the cost of the handover may very likely be higher than the gain. The main problem we address is how location information can be used to guide the selection between the ubiquitous cellular network and any locally available WiFi networks. A main assumption in this section is the availability of a fingerprinting database that contains the average throughput of all available networks for the considered geographical area. In the following, we consider a few different handover decision algorithms that are described in detail in (WHERE D3.7 2010). Here, we give a short overview of the algorithms:

- *LTE only:* This algorithm always connects to the nearest LTE base station, which happens in a seamless manner. This algorithm serves as a comparison for the result of the other algorithms, in the sense that the throughput of the other algorithms should never go below the throughput of this algorithm. If that is the case, no advantage is achieved by making handover to WiFi.
- *Optimal with no handover delay:* This algorithm connects to the network with the highest throughput at any given time. As each handover to or from a WiFi access point has a cost of a 2–3 s delay (WHERE D3.7 2010), the handovers determined by this algorithm are not necessarily the optimal choices. However, due to its simplicity, this algorithm is suitable for being run online.
- *Location-based heuristic prediction algorithm:* This is a heuristic online algorithm that uses previous movement history, location information, and a fingerprinting map for predicting the expected throughput at future positions. This algorithm estimates the future positions of the mobile device for a fixed number of future time-steps by linear extrapolation of the 10 last estimated positions. For each of these future positions, the algorithm creates a list that specifies which network has the highest expected throughput at each time step. Starting from the current time step, the algorithm looks ahead and calculates the expected throughput for each sequence of consecutive and identical entries in the list. If the algorithm reaches a sequence of identical entries where the gain of doing a handover (with 3 s initial handover delay) exceeds the throughput it would achieve if no handover is performed, the algorithm schedules the handover and waits until this handover has been performed. Based on this the algorithm determines when a handover should be performed.
- *Genetic algorithm:* In order to evaluate the goodness of the different algorithms, it is interesting to compare their performance to the optimal handover decision. Since this is not practically feasible to compute due to complexity, (WHERE D3.7 2010) proposed this algorithm, which iteratively tries to find the best choices of the handover, while considering the 2–3 s handover delay. This algorithm represents an estimate of the optimal handover sequence in each scenario, but it is not intended for online use.

9.3.2   Scenario

The proposed algorithms are evaluated in an LTE cellular environment with a hexagonal cell grid and a cell radius of 500 m. In addition, WiFi hotspots are placed randomly in the scenario with coverage range of 100 m. For simplicity in the mobility prediction used in the proposed heuristic algorithm, the WiFi hotspots are placed so that their entire coverage area is within a rectangular area of 1400×900 m.

The movement is modeled with a modified random waypoint model (WHERE D3.7 2010), where waypoints are placed randomly along the edge of the rectangular area. The mobile user will therefore never change direction within WiFi coverage, which is assumed to simplify the mobility prediction. At this point we are interested in assessing the performance of handover decision algorithms and therefore we use a very simple and predictable mobility model.

As mentioned, the entire environment is covered by LTE cells and WiFi hotspots. Rate adaption schemes are assumed for both WiFi and LTE. For details see (WHERE D3.7 2010).

We assume that the delay for handover between LTE cells is insignificant compared to the delay when handing over to or from WiFi, so we disregard it in the following. The expected handover procedure from LTE to WiFi or between two WiFi networks is described in the following.

To perform a handover to a WiFi network, the following actions are taken:

1. *Discovery of networks through scanning:* The discovery of networks typically takes around 0.5 s. This is due to a timeout, since a mobile device waits a certain time before proceeding to the next step to ensure that all relevant WiFi networks are detected.
2. *Authentication:* Takes negligible time.
3. *Association* (for steps 1–3 see [IEEE 802.11 2007] for details): Takes negligible time.
4. *IP obtainment through DHCP* (see [Perkins and Jagannadh 1995] for details): The time spent for the DHCP session is unknown, but empirical tests indicate that it easily takes up to 2 s, and almost never less than 1 s.
5. Mobile *internet protocol (IP)* registration (see [Perkins 1998] for details): If mobile IP is used, it is necessary to register the obtained IP address with the home agent. This can in principle take some time if the round-trip time to the home agent is large. We will assume that the delay due to mobile IP registration is 0.5 s. Of course, in cases where mobile IP is not used, this delay can be disregarded.

Thus, in total the estimated hand over delay varies uniformly between 2 and 3 s.

9.3.3   Summary of Results

(WHERE D3.7 2010) considered the problem of determining when and how a mobile terminal should handover between LTE and WiFi networks in scenarios with ubiquitous LTE coverage and randomly deployed WiFi hotspots. As the task of finding the optimal sequence of handovers that result in the optimal throughput is very complex for realistic scenarios, we have proposed two location-based algorithms that are based on simplifying assumptions and therefore are light enough to be run online. These algorithms rely on positioning information and a map of LTE base stations and WiFi access points to work. We have compared these algorithms to the LTE only case and the estimated optimal result, which was obtained by use of a genetic algorithm. The algorithms were evaluated under varying conditions in terms of WiFi access point deployment density, positioning error, and movement speed.

The results showed that a high movement speed has a negative impact on the performance of the proposed algorithms, due to the potential benefit of performing a handover becoming small when less time is spent in each WiFi hotspot. Also we saw that increasing the access point density was initially beneficial, due to more options for achieving a higher throughput, however, at some point the big selection of access points to choose from seemed to confuse the simplified online algorithms. Finally, we showed that position based handover makes sense when speed is not too high, and positioning error is below 10 m standard deviation. Given that the proposed heuristic prediction algorithm does not cope well with high access point densities and high movement speeds, an obvious future work item would be to make this algorithm more robust. One possible improvement would be to not let the algorithm plan a handover too far into the future without reevaluating the expected benefit as time progresses.

In this section we have considered the time it takes to handover from LTE to WiFi or back to be around 2–3 s, which is a result of a series of actions related to association and IP address obtainment. If this process was optimized, for example, by early preparation of the handover through the network, which would be possible given the availability of positioning information, the benefit of location-based handover enhancement could be further increased.

9.4   Emergency Calls

Important application, use case, and main enabler for the integration of LBS support functions is emergency calls. In September 2011, the European Commission adopted the (EC 2011) recommendation for electronic emergency calls. However, this recommendation does not define clear accuracy requirements for location determination of emergency calls. Earlier, the European Commission has published a recommendation (EC 2003) to follow the recommendations of the *Coordination Group on Access to Location Information for Emergency Services (CGALIES)*. Nevertheless, CGALIES has developed a 'Report on Implementation Issues Related to Access to Location Information by Emergency Services (E-112) in the European Union' (CGALIES 2002):

In the emergency call scenario, the following three tasks have to be executed to make a successful call (CGALIES 2002), where the first two tasks require the same location accuracies in different environments and the third task requires much higher accuracies:

- Call routing and dispatching of emergency request to relevant service and station: The mobile call first needs to be rooted to the relevant service and station. To do so, the Cell-ID and sector ID from the cellular network should be sufficient. As the cells in an urban environment are approximately 1 km large, the required accuracy is 1 km. Similar, in a suburban and rural area, the cell size increase to approximately 10 km and 35 km and so do the accuracies;
- Location of caller or incident:
  - Here, the caller needs to provide location information about the place of emergency. In addition to verbally communicating the approximate or exact location, the cell phone of the caller could provide the emergency service its position. If the caller is able to verbally communicate its position, the cell phone position can be used to verify the callers location

information. In this case the accuracy requirements are between 10–50 m, 30–100 m, 50–100 m, and 20–100 m for indoor and urban, suburban, rural, and highway crossroads. If the caller cannot provide any location information, the requirements for suburban, rural, and highway crossroads change to 10–100 m position accuracy;

– Beside the position accuracy, the time to obtain the position is also relevant and should be available within 30 s of call initiation. An initial position with a accuracy of 200–300 m should be even available within 7 s;

– Additionally, it maybe that several callers refer to the same emergency call. The position accuracy of such a call cluster should be approximately 150 m in urban and 500 m in suburban and rural environments;

– These position accuracies all relate to horizontal accuracies. For vertical accuracies, the position of the cell phone should be known within 10–15 m, that is, 3–4 floors of a building;

– Further, the position estimate of the cell phone should indicate a reliability to the emergency service, such as an CEP of 67% at the lower end and 95% at the upper end;

– In order to map the location of the caller to a local map by the emergency service a map accuracy of 10 m may be required.

Alternative to the location requirements suggested by (CGALIES 2002), the US FCC put up *requirements* for *E911* in terms of location accuracies and reliabilities (FCC 1999) for the operators in the USA. For network based positioning solutions, a position accuracy of at least 100 m should be achieved in 67% of the emergency calls and of at least 300 m in 95% of the emergency calls. In case of handset based positioning solutions, the accuracies should increase to 50 m for 67% of the calls and 150 m to 95% of the calls.

9.5   Location-Based Services–LBS

Almost 30% of the global mobile subscriber bases, around 1.3 billion users, are expected to be using LBS by 2013 (GRAMMAR 2009c). Total mobile search revenues are forecast to reach €3.6 billion by 2013.

The quality of the user experience will be vital in ensuring users adopt these services, and the quality (reliability, availability and accuracy) of the location data is a key factor.

Mobile operators and application developers are coming up with attractive LBS offerings. Terminal vendors are also demonstrating a keen interest in introducing their own location-centric applications and services.

Predicting so-called killer applications for technologies is extremely difficult; however, a number of application areas or topics are likely to be important for location-enabled mobile devices:

• Mobile and location aware advertising
• Social networks
• Navigation and route planning
• Mobile gaming
• Disruptive applications.

9.5.1   Mobile and Location Aware Advertising

Mobile and location aware advertising provides new ways for companies to advertise their products to consumers at or near shopping facilities (Grill 2008). For instance, if the consumer provides its location information to an advertisement server, the system can decide to deliver advertisements based on the consumer's current location or a history of past locations combined with the consumer's profile of interests and purchasing history. The location aware advertising may provide information about nearby shops, offers, and coupons.

Uses cases for offers and coupons may consider individual, friends, loyalty, or charity deals (Puscher 2011). The social network Facebook is currently establishing a LBS based on these four types of deals. For instance, the individual offer of a cinema chain providing a free bag of popcorn when visiting the cinema and checking-in to Facebook. For the friends deal, several friends of a social network need to check-in concurrently at the same location to obtain an offer such as a free bottle of wine. The loyalty deal will reward a consumer for shopping several times at the same facility. In contrast, the charity deal does not reward the consumer directly, but donates the reward to charity.

Besides Facebook, other similar services are provided through Foursquare or Google, for example, Google+ and Google places.

Technology-wise, these services are using the location of the consumer's cell phone obtained by GPS or from a database of cell-IDs and WLAN RSSs (Grill 2008). The critical aspect here is that the consumer's location is obtained in real time at a reasonable accuracy, reliability, and price in terms of used resources, for example, required communication, battery consumption, and so on. Particularly for consumers in urban canyons or indoors, this remains a challenge. Furthermore to avoid a large overhead in communication, zone aware advertisements can be employed.

Another option is to use short range communications such as Bluetooth for location aware advertisement (Ad2Hand 2011). Bluetooth advertisements are made practical by the growing density of Bluetooth in Western countries. In all of Western Europe, almost 100% of smartphones are Bluetooth-enabled. When using proximity marketing via Bluetooth, it is possible to target potential customers in a local area like a shop or conference center. As nearly every person has a cell phone and most of them are Bluetooth enabled, the advertisers can send out calendar events, address information, images, audio, or event video. With a range of 100 m it is possible to scan for Bluetooth enabled cell phones within proximity of the Ad2Hand device. Then depending on what the advertisers are trying to achieve with their campaigns, they can send a small text voucher to bring the customers closer or supply customers with images, audio, or video to sell their products or services.

9.5.2   Social Networks

Social software such as Facebook and Google+ will drive new uses for location information. However, many of these applications also require a communications link–hence the presence of a cellular receiver. Google's Latitude product marketing tag line is 'See where your friends are right now'.

- Latitude, and now its successor Google+, is a new feature for Google Maps on your mobile device. It's also an iGoogle gadget on your computer. Once you've opted in to Latitude, you can see the approximate location of your friends and loved ones who have decided to share their location with you. So now you can do things like see if your spouse is stuck in traffic on the way home from work, notice that a buddy is in town for the weekend, or take comfort in knowing that a loved one's flight landed safely, despite bad weather.

- Map My Tracks:[4] Map My Tracks turns your mobile cell phone into your personal real-time GPS tracking device. Using a mobile cell phone with built-in GPS or an external GPS receiver you can map and track your location in real-time. Map My Tracks is great for bringing a new insight into your sporting activity; it provides a competitive edge and makes training fun. Map My Tracks let your friends, competitors or parents know where you are right now.

9.5.3 Navigation and Route Planning

- Apple, Google, and Microsoft have introduced the option of taking traffic conditions into account in their mapping applications, using traffic data to calculate travel time rather than distance.
- My Location shows your current location on the map, usually within 1000 m, so that you can find out where you are even without GPS. Google Maps for mobile also supports built-in GPS, or can link to a Bluetooth GPS sensor to more accurately pinpoint a user's location. My Location works by recognizing information broadcast from mobile towers near you.
- Transit: Check public transport schedules, determine what transfers you need to make, and plan adventures in more than 50 cities around the world.
- Öffi: An Android App that tells you where and when trains and busses go, including delays and replacement bus service for more and more transport authorities in Europe and beyond.
- Other example applications include: Multimap's Storefinder4Mobile, which is now a part of Microsoft's Bing maps, application puts 'traditional' web based Storefinder search into consumers' pockets: 'Where's my nearest car dealer? Property for sale? WiFi Hotspot?'

9.5.4 Mobile Gaming

Handset developers are addressing the opportunity in creating games that can include a location element such as scavenger hunts or virtual racetracks. Mobile gaming is still in its early stages, but the success of so-called casual gaming have made a huge seller of the Nintendo Wii. Location is seen as the next logical step for gaming. Games currently use touch screens and accelerometers as user controls. The use of embedded GNSS is an obvious extension to this.

9.5.5 Disruptive Applications

Predicting the future of technology is always made difficult because of disruptive technologies that appear to come out of nowhere and gain significant market share. By their very nature disruptive technologies are almost impossible to predict with any level of confidence. Geotagging is an interesting area that may attract increased consumer interest. Geotagging lets users tag

---

[4] From Brent MO 2013 GPS tracking software. [Online; accessed August 23, 2013]

information such as digital photos with location information. While in itself this may not be revolutionary, geotagging may be a technology enabler that drives future applications. Where On Earth (WOE) IDs are already stored by Flickr for uploaded photographs.

IMS Research forecasts that the GPS Camera market will grow to over 40 million units in 2012, across both professional and consumer orientated cameras.

One company taking an innovative approach to Geotagging is Geotate (www.geotate.com/). Geotate's software automatically adds location labels to photos by assigning raw (unprocessed GPS signals) to the image file which is then post processed to derive the location information.

Geotate automatic geotagging technology uses a 0.2 s GPS acquisition signal to capture a RAW GPS signal as part of the usual photo moment, ensuring no delays for the user. In comparison, a traditional streaming GPS device will take a few seconds to find the satellite signals, around 30 s to download satellite orbit information, and a few more seconds to calculate the user's position. In 2009, Geotate has been acquired by u-blox AG, a Swiss based GPS mass market receiver manufacturer.

9.5.6   Future Applications

There could be an entirely new breed of applications coming into future cell phone handsets, possibly with the iPhone–driven by Apple's iTunes Applications Store. While Apple has taken the lead in this area, we now see other vendors following. While many of the estimated 900 000+ applications in the iTunes Apps Store do not exploit the GNSS capabilities of the iPhone there is great potential for user generated applications in (Iskold 2008):

1. Reality tagging–Reality tagging will be like a distributed Google Earth, but for pictures. One can take a picture of landmark, then comment and add tags. The phone will automatically geotag it and sends the picture to a photo sharing service on the Web.
2. People tagging–Using cell phones to tag people.
3. Reality recognition–Reality recognition will be fuelled by reality tagging and advanced image recognition.
4. Physical social networks–One can walk into a restaurant, open up the cell phone and find a list of friends who are around.
5. Personalized travel guides–People love tours and tour guides who tell them about landmarks and history of new places. This application could have personalized tours of any location of the world available. Information about the current location could appear in the hand with a touch of a button.
6. Distributed mobile games–We can expect to see interesting activity in mobile games and we will be able to play games that take place in both the physical and digital worlds.
7. Physical browsing and digital shopping–Browsing things in the physical world and instantly buying them online. For example, it will be possible to take a picture of a barcode on a book in the store and have it immediately appear in the shopping cart.
8. Location/time based deals–If a person is standing in front of a store, for example, to buy a camera, pointing the phone to a camera could trigger instant bids from neighborhood stores.

Out of all these predicted applications for a cell phone, the majority of applications need accurate location information. These applications may not be the 'killer application' that drives users to demand positioning capability in their handsets, however, they do expose the wireless location and GNSS functionality beyond simple navigation applications. Therefore, there is increasing demand for wireless location and GNSS based applications in future handsets as the popularity and value of these applications increases.

References

ETS 2009 *3GPP-LTE; Evolved Universal Terrestrial Radio Access (E-UTRA); Physical layer procedures.* ETSI TS 136 213 V8.8.0 (2009-06).

3GPP 2013 3GPP work programme. [Online; accessed June 21, 2013].

3GPP M 2001 Overview of 2G LCS technologies and standards *3GPP TSG SA2 LCS Workshop*, London, UK.

3GPP S 2010 *RP-101425: Coordinated Multi-Point Operation for LTE* 3rd Generation Partnership Project (3GPP), Samsung.

3GPP, TSGRAN 2013 *Location Measurement Unit (LMU) performance specification; Network Based Positioning Systems in E-UTRAN* 3GPP. 3GPP TS 36.111 V1.1.1.

3GPP2 2004 *Position Determination Service for cdma2000 Spread Spectrum Systems* 3rd Generation Partnership Project 2 (3GPP2). C.S0022-A v1.0.

Ad2Hand 2011 *Bluetooth advertising with a free software solution* www.ad2hand.co.uk/ [Online; accessed November 18, 2013].

AeroScout 2013 Aeroscout system technology overview www.aeroscout.com/technology [Online; accessed November 18, 2013].

Alamouti SM 1998 A simple transmit diversity technique for wireless communications. *IEEE Journal on Selected Areas in Communications* **16**(8), 1451–1458.

Ali-Löytty S 2009 *Gaussian Mixture Filters in Hybrid Positioning* PhD thesis Tampere University of Technology. http://URN.fi/URN:NBN:fi:tty-200905191055 [Online; accessed November 18, 2013].

Alippi C, Cogliati D and Vanini G 2006 A statistical approach to localize passive RFIDs. *IEEE International Symposium on Circuits and Systems (ISCAS) 2006*.

Apple 2013 iOS3 6: Understanding location services http://support.apple.com/kb/HT5467 [Online; accessed November 18, 2013].

Arulampalam MS, Maskell S, Gordon N and Clapp T 2002 A tutorial on particle filters for online nonlinear/non-Gaussian Bayesian tracking. *IEEE Transactions on Signal Processing* **50**(2), 174–188.

Astely D, Dahlman E, Furuskär A, Jading Y, Lindström M and Parkvall S 2009 LTE: The evolution of mobile broadband. *IEEE Communications Magazine* **47**(4), 44–51.

Atmel 2013 *Data Sheet Atmel AT86RF233 – Low Power, 2.4GHz Transceiver for ZigBee, RF4CE, IEEE 802.15.4, 6LoWPAN, and ISM Applications* Atmel.

Bahl P and Padmanabhan VN 2000 Radar: An in-building RF-based user location and tracking system. *Proceedings of the IEEE International Conference on Computer Communications (IEEE INFOCOM 2000)*, vol. 2, pp. 775–785, Tel-Aviv, Israel.

Positioning in Wireless Communications Systems, First Edition. Stephan Sand, Armin Dammann and Christian Mensing.
© 2014 John Wiley & Sons, Ltd. Published 2014 by John Wiley & Sons, Ltd.

Baniukevic A, Sabonis D, Jensen CS and Lu H 2011 Improving Wi-Fi based indoor positioning using Bluetooth add-ons. *2011 12th IEEE International Conference on Mobile Data Management (MDM)*, vol. 1, pp. 246–255.

Bar-Shalom Y, Li RX and Kirubarajan T 2001 *Estimation with Applications to Tracking and Navigation: Theory Algorithms and Software*. John Wiley & Sons, Ltd, Chichester, Uk.

Bedford M and Kennedy G 2012 Evaluation of ZigBee (IEEE 802.15.4) time-of-flight-based distance measurement for application in emergency underground navigation. *IEEE Transactions on Antennas and Propagation* **60**(5), 2502–2510.

Berggren F and Popovic BM 2007 A non-hierarchical cell search scheme. *IEEE Wireless Communications and Networking Conference (WCNC)*.

Bittins B and Sieck J 2012 Multimodal and collaborative localisation service for diverse environments. *IEEE 1st International Symposium on Wireless Systems (IDAACS-SWS) 2012*, pp. 28–33.

Bluetooth 2013 A look at the basics of Bluetooth wireless technology www.bluetooth.com/Pages/Basics.aspx [Online; accessed November 18, 2013].

Bluetooth SIG 2010 *Specification of the Bluetooth System* Bluetooth Special Interest Group. Core specification version 4.0.

Briers M, Maskell S and Wright R 2003 A Rao–Blackwellized unscented Kalman filter. *Proceedings of the International Conference on Information Fusion* **1**, 55–61.

Cardinali R, De Nardis L, Di Benedetto M and Lombardo P 2006 UWB ranging accuracy in high- and low-data-rate applications. *IEEE Transactions on Microwave Theory and Techniques* **54**(4), 1865–1875.

CGALIES 2002 *Final Report: Report on Implementation Issues Related to Access to Location Information by Emergency Services (E112) in the European Union* Coordination Group on Access to Location Information for Emergency Services (CGALIS).

Chan FWC and So HC 2009 Accurate Distributed Range-Based Positioning Algorithm for Wireless Sensor Networks. *IEEE Transactions on Signal Processing* **57**(10), 4100–4105.

Chen PC 1999 A non-line-of-sight error mitigation algorithm in location estimation. *Proceedings of the IEEE Wireless Communications and Networking Conference (WCNC)* **1**, 316–320.

Cherntanomwong P and Suroso D 2011 Indoor localization system using wireless sensor networks for stationary and moving targets. 8th *International Conference on Information, Communications and Signal Processing (ICICS) 2011*, pp. 1–5.

Cisco 2012 Cisco visual networking index: Global mobile data traffic forecast update, 2011-2016. white paper.

Ciurana M, Cugno S and Barcel-Arroyo F 2007 WLAN indoor positioning based on TOA with two reference points *Proceedings of the 4th Workshop On Positioning, Navigation and Communication (WPNC 2007)*, pp. 23–28, Hanover, Germany.

Cong L and Zhuang W 2005 Nonline-of-sight error mitigation in mobile location. *IEEE Transactions on Wireless Communications* **4**(2), 560–573.

Dammann A and Kaiser S 2001 Standard conformable antenna diversity techniques for OFDM and its application to the DVB-T system. *Proceedings IEEE Global Telecommunications Conference (GLOBECOM 2001)*, San Antonio, TX, USA, pp. 3100–3105.

Dardari D, Conti A, Ferner U, Giorgetti A and Win M 2009 Ranging with ultrawide bandwidth signals in multipath environments. *Proceedings of the IEEE* **97**(2), 404–426.

DecaWave 2013 ScenSor advance product information DW1000 www.decawave.com/scensor.html [Online; accessed November 18, 2013].

di Flora C and Hermersdorf M 2008 A practical implementation of indoor location-based services using simple WiFi. *Journal of Location Based Services* **2**(2), 87–111.

Djuric PM, Kotecha JH, Zhang J, Huang Y, Ghirmai T, Bugallo MF and Miguez J 2003 Particle filtering. *IEEE Signal Processing Magazine* pp. 19–38.

Doucet A, de Freitas JFG and Gordon NJ 2001 An introduction to sequential Monte Carlo methods In *Sequential Monte Carlo Methods in Practice* (eds Doucet A, de Freitas JFG and Gordon NJ). Springer Verlag, Heidelberg, Germany.

EC 2003 *Commission Recommendation of 25 July 2003 on the processing of caller location information in electronic communication networks for the purpose of location-enhanced emergency call services* European Comission (EC). 2003/559/EC.

EC 2011 *Commission recommendation of 8 September 2011 on support for an EU-wide eCall service in electronic communication networks for the transmission of in-vehicle emergency calls based on 112 (eCalls)* European Comission (EC). 2011/750/EC.

ECMA 2008 *Standard ECMA-368 – High Rate Ultra Wideband PHY and MAC Standard* 3rd edn ECMA International.

ECMA 2010 *Standard ECMA-387 – High Rate 60 GHz PHY, MAC and PALs* 2nd edn. ECMA International.

ECMA 2013a *Standard ECMA-340 – Near Field Communication–Interface and Protocol (NFCIP-1)* 3rd edn ECMA International.

ECMA 2013b *Standard ECMA-352 – Near Field Communication–Interface and Protocol (NFCIP-2)* 3rd edn ECMA International.

EDA n.d. EGNOS Data Access Service. www.insidegnss.com [Online; accessed November 18, 2013].

EMS n.d. EGNOS Message Server. www.egnos-pro.esa.int/ems [Online; accessed November 18, 2013].

EPC HF RFID 2011 *EPC^{TM} Radio-Frequency Identity Protocols – EPC Class-1 HF RFID Air Interface Protocol for Communications at 13.56 MHz, Version 2.0.3* GS1 EPCglobal.

EPC UHF RFID 2008 *EPC^{TM} Radio-Frequency Identity Protocols – EPC Class-1 Generation-2 UHF RFID Protocol for Communications at 860 MHz–960 MHz, Version 1.2.0* GS1 EPCglobal.

ESA 2007 Galileo System Simulation Facility (GSSF) www.gssf.info [Online; accessed November 18, 2013].

ESA 2011 EGNOS navigation system begins serving Europe's aircraft www.esa.int/Our_Activities/Technology/TTP2/EGNOS_navigation_system_begins_serving_Europe_s_aircraft [Online; accessed November 18, 2013].

Etemad K 2008 Overview of mobile WiMAX technology and evolution. *IEEE Communications Magazine* **46**(10), 31–40.

ETS 2013a *3GPP-LTE; Evolved Universal Terrestrial Radio Access (E-UTRA); Physical layer; Measurements*. ETSI TS 136 214 V11.1.0 (2013-02).

ETS 2013b *LTE; Evolved Universal Terrestrial Radio Access (E-UTRA) and Evolved Universal Terrestrial Radio Access Network (E-UTRAN); Overall description; Stage 2*. ETSI TS 136 300 V11.5.0 (2013-04).

ETS 2013c *LTE; Evolved Universal Terrestrial Radio Access (E-UTRA); LTE Positioning Protocol A (LPPa)*. ETSI TS 136 455 V11.1.0 (2013-01).

ETS 2013d *LTE; Evolved Universal Terrestrial Radio Access (E-UTRA); LTE Positioning Protocol (LPP)*. ETSI TS 136 355 V11.2.0 (2013-04).

ETS 2013e *LTE; Evolved Universal Terrestrial Radio Access Network (E-UTRAN); Stage 2 functional specification of User Equipment (UE) positioning in E-UTRAN*. ETSI TS 136 305 V11.3.0 (2013-04).

ETSI 2003 *Near Field Communication (NFC) IP-1; Interface and Protocol (NFCIP-1)* ETSI. ETSI TS 102 190 V1.1.1 (2003-03).

ETSI 2004 *Near Field Communication Interface and Protocol-2 (NFCIP-2)* ETSI. ETSI TS 102 312 V1.1.1 (2004-02).

ETSI 2006 *Telecommunications and Internet converged Services and Protocols for Advanced Networking (TISPAN); Overview of Radio Frequency Identification (RFID) Tags in the telecommunications industry* European Telecommunications Standard Institute (ETSI). ETSI TR 102 449 V1.1.1.

ETSI 2010 *Universal Mobile Telecommunications System (UMTS); UTRAN functions, examples on signalling procedures* European Telecommunications Standard Institute (ETSI). ETSI TS 125 931 V9.0.0.

ETSI 2012a *Digital cellular telecommunications system (Phase 2+); Functional stage 2 description of Location Services (LCS) in GERAN* European Telecommunications Standard Institute (ETSI). ETSI TS 143 059 V11.0.0.

ETSI 2012b *Digital cellular telecommunications system (Phase 2+); Location Services (LCS); Mobile Station (MS)–Serving Mobile Location Centre (SMLC) Radio Resource LCS Protocol (RRLP)* European Telecommunications Standard Institute (ETSI). ETSI TS 144 031 V11.0.0.

ETSI 2012c *Digital cellular telecommunications system (Phase 2+); Modulation* European Telecommunications Standard Institute (ETSI). ETSI TS 145 004 V11.0.0.

ETSI 2012d *Digital cellular telecommunications system (Phase 2+); Physical layer on the radio path; General description* European Telecommunications Standard Institute (ETSI). ETSI TS 145 001 V11.0.0.

ETSI 2012e *Digital cellular telecommunications system (Phase 2+); Radio subsystem link control* European Telecommunications Standard Institute (ETSI). ETSI TS 145 008 V11.1.0.

ETSI 2012f *LTE; Requirements for further advancements for Evolved Universal Terrestrial Radio Access (E-UTRA) (LTE-Advanced)* ETSI. ETSI TR 136 913 V11.0.0 (2012-11).

ETSI 2012g *Universal Mobile Telecommunications System (UMTS); LTE; Requirements for Evolved UTRA (E-UTRA) and Evolved UTRAN (E-UTRAN)* ETSI. ETSI TR 125 913 V9.0.0 (2010-02).

ETSI 2012h *Universal Mobile Telecommunications System (UMTS); Physical layer; Measurements (FDD)* European Telecommunications Standard Institute (ETSI). ETSI TS 125 215 V11.0.0.

ETSI 2012i *Universal Mobile Telecommunications System (UMTS); Physical layer; Measurements (TDD)* European Telecommunications Standard Institute (ETSI). ETSI TS 125 225 V11.0.0.

ETSI 2012j *Universal Mobile Telecommunications System (UMTS); Stage 2 functional specification of User Equipment (UE) positioning in UTRAN* European Telecommunications Standard Institute (ETSI). ETSI TS 125 305 V11.0.0.

ETSI 2013a *Digital cellular telecommunications system (Phase 2+); Radio transmission and reception* European Telecommunications Standard Institute (ETSI). ETSI TS 145 005 V11.2.0.

ETSI 2013b *LTE; Evolved Universal Terrestrial Radio Access (E-UTRA); Base Station (BS) radio transmission and reception* ETSI. ETSI TS 136 104 V11.3.1.

ETSI 2013c *LTE; Evolved Universal Terrestrial Radio Access (E-UTRA); Long Term Evolution (LTE) physical layer; General description* ETSI. ETSI TS 136 201 V11.1.0.

ETSI 2013d *LTE; Evolved Universal Terrestrial Radio Access (E-UTRA); Multiplexing and channel coding* ETSI. ETSI TS 136 212 V11.2.0.

ETSI 2013e *LTE; Evolved Universal Terrestrial Radio Access (E-UTRA); Physical channels and modulation* ETSI. ETSI TS 136 211 V11.1.0.

ETSI 2013f *LTE; Evolved Universal Terrestrial Radio Access (E-UTRA); Requirements for support of radio resource management* ETSI. ETSI TS 136 113 V11.4.0.

ETSI 2013g *LTE; Evolved Universal Terrestrial Radio Access (E-UTRA); SLM interface Application Protocol (SLmAP)* ETSI. ETSI TS 136 459 V11.1.0.

ETSI 2013h *LTE; Evolved Universal Terrestrial Radio Access (E-UTRA); User Equipment (UE) radio transmission and reception* ETSI. ETSI TS 136 101 V11.4.0.

ETSI 2013i *Universal Mobile Telecommunications System (UMTS); Physical layer–general description* European Telecommunications Standard Institute (ETSI). ETSI TS 125 201 V11.1.0.

ETSI 2013j *Universal Mobile Telecommunications System (UMTS); Requirements for support of radio resource management (FDD)* European Telecommunications Standard Institute (ETSI). ETSI TS 125 133 V11.4.0.

ETSI 2013k *Universal Mobile Telecommunications System (UMTS); Requirements for support of radio resource management (TDD)* European Telecommunications Standard Institute (ETSI). ETSI TS 125 123 V11.3.0.

ETSI 2013l *Universal Mobile Telecommunications System (UMTS); User Equipment (UE) radio transmission and reception (FDD)* European Telecommunications Standard Institute (ETSI). ETSI TS 125 101 V11.4.0.

Fang SH, Wang CH, Huang TY, Yang CH and Chen YS 2012 An enhanced ZigBee indoor positioning system with an ensemble approach. *IEEE Communications Letters* **16**(4), 564–567.

Fazel K and Kaiser S 2003 *Multi-Carrier and Spread Spectrum Systems*. John Wiley & Sons, Ltd, Chichester, UK. ISBN 0-470-84899-5.

FCC 1999 *Revision of the Commission's Rules to Ensure Compatibility with Enhanced 911 Emergency Calling Systems: Third Report and Order* Federal Communications Commission (FCC).

FCC 2002 *Revision of Part 15 of the Commission's Rules Regarding Ultra-Wideband Transmission Systems: First Report and Order* Federal Communications Commission (FCC).

Figueiras J and Frattasi S 2010 *Mobile Positioning and Tracking: From Conventional to Cooperative Techniques*. John Wiley & Sons, Ltd, Chichester, UK.

Fontana R 2004 Recent system applications of short-pulse ultra-wideband (UWB) technology. *IEEE Transactions on Microwave Theory and Techniques* **52**(9), 2087–2104.

Forno F, Malnati G and Portelli G 2005 Design and implementation of a Bluetooth ad hoc network for indoor positioning. *IEE Proceedings–Software* **152**(5), 223–228.

Foy W 1976 Position-location solutions by Taylor-series estimation. *IEEE Transactions on Aerospace and Electronic Systems* **AES-12**(2), 187–194.

Frattasi S 2007 *Link Layer Techniques Enabling Cooperation in Fourth Generation Wireless Networks*. PhD thesis Aalborg University Aalborg, Denmark.

Gal 2010 *European GNSS (Galileo) Open Service–Signal In Space Interface Control Document, Issue 1.1*.

Gezici S, Tian Z, Giannakis GB, Kobayashi H, Molisch AF, Poor HV and Sahinoglu Z 2005 Localization via ultra-wideband radios. *IEEE Signal Processing Magazine* **22**(4), 70–84.

GLO 2002 *GLONASS Interface Control Document*. Version 5.0.

Godara L 1997a Applications of antenna arrays to mobile communications, part i. *Proceedings of the IEEE* **85**(8), 1195–1245.

Godara L 1997b Applications of antenna arrays to mobile communications, part ii. *Proceedings of the IEEE* **85**(8), 1195–1245.

Gold R 1967 Optimal binary sequences for spread spectrum multiplexing. *IEEE Transactions on Information Theory* **13**(4), 619–621.

Golden SA and Bateman SS 2007 Sensor measurements for Wi-Fi location with emphasis on time-of-arrival ranging. *IEEE Transactions on Mobile Computing* **6**(10), 1185–1198.

Google 2013 Location source and accuracy https://support.google.com/gmm/answer/81873 [Online; accessed November 18, 2013].

Gordon N, Salmond D and Smith AFM 1993 Novel approach to non-linear and non-Gaussian Bayesian state estimation. *Proceedings of the IEE* **140**, 107–113.

GRAMMAR 2009a *Gap Analysis of GNSS receivers and technology* Galileo Ready Advanced Mass Market Receiver (GRAMMAR) consortium.

GRAMMAR 2009b GRAMMAR Website. www.gsa-grammar.eu [Online; accessed November 18, 2013].

GRAMMAR 2009c *Market Definition and Core Technology Report* Galileo Ready Advanced Mass Market Receiver (GRAMMAR) consortium.

Grill A 2008 *Location Based Advertising–An Introduction* http://londoncalling.co/2008/05/location-based-advertising-introduction/ [Online; accessed November 18, 2013].

Gustafsson F and Gunnarsson F 2005 Mobile positioning using wireless networks. *IEEE Signal Processing Magazine* **22**(4), 41–53.

Gustafsson F, Gunnarsson F, Bergman N, Forssell U, Jansson J, Karlsson R and Nordlund PJ 2002 Particle filters for positioning, navigation, and tracking. *IEEE Transactions on Signal Processing* **50**(2), 425–437.

Guvenc I and Chong CC 2009 A survey on TOA based wireless localization and NLOS mitigation techniques. *IEEE Communications Surveys & Tutorials* **11**(3), 107–124.

Guvenc I, Gezici S and Sahinoglu Z 2010 Fundamental limits and improved algorithms for linear least-squares wireless position estimation. *Wireless Communications and Mobile Computing* **12**, 1037–1052.

Hallberg J, Nilsson M and Synnes K 2003 Positioning with Bluetooth 10th *International Conference on Telecommunications (ICT) 2003*, vol. **2**, pp. 954–958.

Hammadi O, Hebsi A, Zemerly M and Ng J 2012 Indoor localization and guidance using portable smartphones *IEEE/WIC/ACM International Conferences on Web Intelligence and Intelligent Agent Technology (WI-IAT) 2012*, vol. **3**, pp. 337–341.

Haykin S 2002 *Adaptive Filter Theory* 4th edn. Prentice-Hall, Upper Saddle River, NJ, USA.

Hermersdorf M 2006 Indoor positioning with a WLAN access point list on a mobile device *Proceedings of the Workshop on World-Sensor-Web (WSW 2006) at 4th ACM Conference on Embedded Networked Sensor Systems (SenSys 2006)*, Boulder, CO, USA.

Hewlett-Packard, Intel, LSI, Microsoft, NEC, Samsung and ST-Ericsson 2010 *Wireless Universal Serial Bus Specification 1.1* USB Implementers Forum.

Hiertz G, Denteneer D, Stibor L, Zang Y, Costa X and Walke B 2010 The IEEE 802.11 universe. *IEEE Communications Magazine* **48**(1), 62–70.

Ho K and Chan Y 1997 Geolocation of a known altitude object from TDOA and FDOA measurements. *IEEE Transactions on Aerospace and Electronic Systems* **33**(3), 770–783.

Hossain A and Soh WS 2007 A comprehensive study of Bluetooth signal parameters for localization *Personal, Indoor and Mobile Radio Communications, 2007. PIMRC 2007. IEEE 18th International Symposium on*, pp. 1–5.

Huawei 2005 *3GPP R1-050841: Further analysis of soft-frequency reuse scheme* 3GPP TSG RAN WG1#42 London.

IEE 2004 *IEEE P802.11 Wireless LANs, TGn Channel Models*. IEEE 802.11-03/940r2.

IEEE 802.11 2007 *IEEE Standard for Information technology – Telecommunications and information exchange between systems – Local and metropolitan area networks – Specific requirements – Part 11: Wireless LAN Medium Access Control (MAC) and Physical Layer (PHY) Specifications* IEEE.

IEEE 802.11 2012 *IEEE Standard for Information technology – Telecommunications and information exchange between systems – Local and metropolitan area networks – Specific requirements – Part 11: Wireless LAN Medium Access Control (MAC) and Physical Layer (PHY) Specifications* IEEE.

IEEE 802.11ad 2012 *IEEE Standard for Information technology – Telecommunications and information exchange between systems – Local and metropolitan area networks – Specific requirements – Part 11: Wireless LAN Medium Access Control (MAC) and Physical Layer (PHY) Specifications –Amendment 3: Enhancements for Very High Throughput in the 60 GHz Band* IEEE.

IEEE 802.11k 2008 *IEEE Standard for Information technology – Telecommunications and information exchange between systems – Local and metropolitan area networks – Specific requirements – Part 11: Wireless LAN Medium Access Control (MAC) and Physical Layer (PHY) Specifications –Amendment 1: Radio Resource Measurement of Wireless LANs* IEEE.

IEEE 802.11u 2011 *IEEE Standard for Information technology – Telecommunications and information exchange between systems – Local and metropolitan area networks – Specific requirements – Part 11: Wireless LAN Medium Access Control (MAC) and Physical Layer (PHY) Specifications –Amendment 9: Interworking with External Networks* IEEE.

IEEE 802.11v 2011 *IEEE Standard for Information technology – Telecommunications and information exchange between systems – Local and metropolitan area networks – Specific requirements – Part 11: Wireless LAN Medium Access Control (MAC) and Physical Layer (PHY) specifications –Amendment 8: IEEE 802.11 Wireless Network Management* IEEE.

IEEE 802.15.3c 2009 *IEEE Standard for Information technology – Telecommunications and information exchange between systems – Local and metropolitan area networks – Specific requirements – Part 15.3: Wireless Medium Access Control (MAC) and Physical Layer (PHY) Specifications for High Rate Wireless Personal Area Networks (WPANs) – Amendment 2: Millimeter-wave-based Alternative Physical Layer Extension* IEEE.

IEEE 802.15.4 2003 *IEEE Standard for Information Technology – Telecommunications and Information Exchange Between Systems – Local and Metropolitan Area Networks – Specific Requirements – Part 15.4: Wireless Medium Access Control (MAC) and Physical Layer (PHY) Specifications for Low-Rate Wireless Personal Area Networks (LR-WPANs)* IEEE.

IEEE 802.15.4 2011 *IEEE Standard for Local and metropolitan area networks – Part 15.4: Low-Rate Wireless Personal Area Networks (LR-WPANs)* IEEE.

IEEE 802.15.4a 2007 *IEEE Standard for Information Technology – Telecommunications and Information Exchange Between Systems – Local and Metropolitan Area Networks – Specific Requirements – Part 15.4: Wireless Medium Access Control (MAC) and Physical Layer (PHY) Specifications for Low-Rate Wireless Personal Area Networks (WPANs) – Amendment 1: Add Alternate PHYs* IEEE.

IEEE 802.15.4f 2012 *IEEE Standard for Local and metropolitan area networks – Part 15.4: Low-Rate Wireless Personal Area Networks (LR-WPANs) – Amendment 2: Active Radio Frequency Identification (RFID) System Physical Layer (PHY)* IEEE.

IEEE 802.16 2012 *IEEE Standard for Air Interface for Broadband Wireless Access Systems, IEEE Std 802.16-2012 (Revision of IEEE Std 802.16-2009)* IEEE.

IEEE 802.16.1 2012 *IEEE Standard for WirelessMAN-Advanced Air Interface for Broadband Wireless Access Systems, IEEE Std 802.16.1-2012* IEEE.

In-Location Alliance 2013 In-Location Alliance – introduction presentation on http://in-location-alliance.com/files/ILA/In-Location_Alliance_Introduction.pdf [Online; accessed November 18, 2013].

infsoft GmbH 2013 Indoor & outdoor positioning http://www.infsoft.com/indoor-positioning-navigation/ [Online; accessed November 18, 2013].

Ingram S, Harmer D and Quinlan M 2004 Ultrawideband indoor positioning systems and their use in emergencies. *Position Location and Navigation Symposium (PLANS) 2004*, pp. 706–715.

Inoue M, Fujii T and Nakagawa M 2002 Space time transmit site diversity for OFDM multi base station system 4th *International Workshop on Mobile and Wireless Communications Network (MWCN) 2002*, pp. 30–34, Stockholm, Sweden.

Iskold A 2008 12 future Apps for your iPhone www.readwriteweb.com/archives/12_future_apps_for_your_iphone.php [Online; accessed November 18, 2013].

ISO/IEC 14443 2008-2012 *ISO/IEC 14443 (all parts): Identification cards – Contactless integrated circuit cards – Proximity cards* ISO/IEC.

ISO/IEC 15693 2008-2010 *ISO/IEC 15693 (all parts): Identification cards – Contactless integrated circuit cards – Vicinity cards* ISO/IEC.

ISO/IEC 18000 2008-2013 *ISO/IEC 18000 (all parts): Information technology – Radio frequency identification for item management* ISO/IEC.

ISO/IEC 18092 2013 *ISO/IEC 18092:2013 Information technology – Telecommunications and information exchange between systems – Near Field Communication – Interface and Protocol (NFCIP-1)* ISO/IEC.

ISO/IEC 21481 2012 *ISO/IEC 21481:2012 Information technology – Telecommunications and information exchange between systems – Near Field Communication – Interface and Protocol -2 (NFCIP-2)* ISO/IEC.

IST-2003-507581 WINNER 2007 WINNER II Deliverable D1.1.2: WINNER II Channel Models.

ITU-R 2006 *RECOMMENDATION ITU-R SM.1755: Characteristics of ultra-wideband technology* International Telecommunication Union Radiocommunication (ITU-R) Sector.

Izquierdo F, Ciurana M, Barceló F, Paradells J and Zola E 2006 Performance evaluation of a TOA-based trilateration method to locate terminals in WLAN *Proceedings of the 1st International Symposium on Wireless Pervasive Computing (ISWPC 2006)*, pp. 1–6, Phuket, Thailand.

JTC 1993 *Final Report on RF Channel Characterization.* JTC(AIR)/93.09.23-238R2.

Julier SJ and Uhlmann JK 1997 A new extension of the Kalman filter to nonlinear systems. *Proceedings of AeroSense* 11th Int. Symp. Aerospace/Defense Sensing, Simulation and Controls. pp. 182–193.

Julier SJ and Uhlmann JK 2004 Unscented filtering and nonlinear estimation. *Proceedings of the IEEE* **92**(3), 401–422.

Kalman RE 1960 A new approach to linear filtering and prediction problems. *Transactions of the ASME-Journal of Basic Engineering* **82**(D), 35–45.

Kammann J, Angermann M and Lami B 2003 A new mobility model based on maps. *Proceedings of the IEEE Vehicular Technology Conference (VTC) Fall.*

Kay SM 1993 *Fundamentals of Statistical Signal Processing: Estimation Theory.* Prentice Hall.

Kim I, Han Y, Kim Y and Bang S 2008 Sequence hopping cell search scheme for OFDM cellular systems. *IEEE Transactions on Wireless Communications* **7**(5), 1483–1489.

King T, Lemelson H, Farber A and Effelsberg W 2009 BluePos: Positioning with Bluetooth. *IEEE International Symposium on Intelligent Signal Processing (WISP) 2009*, pp. 55–60.

Kotanen A, Hannikainen M, Leppakoski H and Hamalainen T 2003 Experiments on local positioning with Bluetooth. *International Conference on Information Technology: Coding and Computing [Computers and Communications] (ITCC) 2003*, pp. 297–303.

Krach B 2010 *Sensor Fusion by Bayesian Filtering for Seamless Pedestrian Navigation.* PhD thesis University of Erlangen-Nürnberg.

Krach B and Weigel R 2009 Markovian channel modeling for multipath mitigation in navigation receivers. *Proceedings of the European Conference on Antennas and Propagation (EuCAP)* pp. 1441–1445.

Krach B, Lentmaier M and Robertson P 2008 Bayesian detection and tracking for joint positioning and multipath mitigation in GNSS. *Proceedings of the Workshop on Positioning, Navigation and Communication (WPNC)* pp. 173–180.

Lee JY and Scholtz R 2002 Ranging in a dense multipath environment using an UWB radio link. *IEEE Journal on Selected Areas in Communications* **20**(9), 1677–1683.

Levenberg K 1944 A method for the solution of certain non-linear problems in least squares. *Quarterly Journal of Applied Mathematics* **2**, 164–168.

Li S, Chen S, Lou Y, Lu B and Liang Y 2012 A recurrent neural network for inter-localization of mobile phones. *The 2012 International Joint Conference on Neural Networks (IJCNN)*, pp. 1–5.

Liu H, Darabi H, Banerjee P and Liu J 2007a Survey of wireless indoor positioning techniques and systems. *IEEE Transactions on Systems, Man, and Cybernetics – Part C: Applications and Reviews* **37**(6), 1067–1080.

Liu P, Tao Z, Narayanan S, Korakis T and Panwar S 2007b CoopMAC: A cooperative MAC for wireless LANs. *IEEE Journal on Selected Areas in Communications* **25**(2), 340.

LOR 1994 *Specification of the transmitted LORAN-C Signal.* COMDTINST M16562.4A.

Maheshwari H and Kemp A 2011 On the enhanced ranging performance for IEEE 802.15.4 compliant WSN devices. *4th IFIP International Conference on New Technologies, Mobility and Security (NTMS) 2011*, pp. 1–5.

Mahtab Hossain A, Jin Y, Soh WS and Van HN 2013 SSD: A robust RF location fingerprint addressing mobile devices' heterogeneity. *IEEE Transactions on Mobile Computing* **12**(1), 65–77.

Marquardt DW 1963 An algorithm for least-squares estimation of nonlinear parameters. *SIAM Journal on Applied Mathematics* **11**(2), 431–441.

Maybeck P 1979 *Stochastic Models, Estimation and Control, Volume I.* Academic Press.

Mayorga CLF, della Rosa F, Wardana SA, Simone G, Raynal MCN, Figueiras J, and Frattasi S 2007 Cooperative Positioning Techniques for Mobile Localization in 4G Cellular Networks. *Proceedings of the IEEE International Conference on Pervasive Services.*

Mensing C 2013 *Location Determination in OFDM Based Mobile Radio Systems.* PhD thesis Technische Universität München.

Mensing C and Nielsen J 2010 Centralized cooperative positioning and tracking with realistic communications constraints. *Proceedings of the Workshop on Positioning, Navigation, and Communication (WPNC).*

Mensing C and Plass S 2006 Positioning algorithms for cellular networks using TDOA. *Proceedings of the IEEE International Conference on Acoustics, Speech, and Signal Processing (ICASSP), Toulouse, France.*

Mensing C, Sand S, Dammann A and Utschick W 2009 Data-aided location estimation in cellular OFDM communications systems. *Proceedings of the IEEE Global Communications Conference (GLOBE-COM), Honolulu, HI, USA.*

Miller LE, Wilson PF, Bryner NP, Harris MH, Guerrieri JR, Stroup DW and Klein-Berndt L 2006 RFID-assisted indoor localization and communication for first responders. *International Symposium on Advance Radio Technologies (ISART)*, Boulder, CO, USA.

Minn H, Bhargava VK and Letaief KB 2003 A robust timing and frequency synchronization for OFDM systems. *IEEE Transactions on Wireless Communications* **2**(4), 822–839.

Musso C, Oudjane N and LeGland F 2001 Improving regularized particle filters In *Sequential Monte Carlo Methods in Practice* (ed. Doucet A, de Freitas JFG and Gordon NJ). Springer Verlag, Heidelberg, Germany.

Nan 2007 Real time location systems www.nanotron.com/EN/SU_docs_white_papers.php [Online; accessed November 18, 2013]. White Paper, Nanotron Technologies GmbH.

NFC Forum 2013 Near field communication forum www.nfc-forum.org/home/ [Online; accessed November 18, 2013].

NXP Semiconductors 2013 *Data Sheet NXP Jennic JN5148-001 – IEEE802.15.4 Wireless Microcontroller* NXP.

OMA 2011 *Enabler Release Definition for Secure User Plane Location (SUPL)* Open Mobile Alliance (OMA). Candidate Version 3.0.

OMA 2013 Open moblie alliance http://openmobilealliance.org/. [Online; accessed November 18, 2013]

Ozdenizci B, Ok K, Coskun V and Aydin M 2011 Development of an indoor navigation system using NFC technology. *Fourth International Conference on Information and Computing (ICIC) 2011*, pp. 11–14.

Parkinson BW and Spilker Jr. JJ 1996 *Global Positioning System: Theory and Applications, Volume 1.* Progress in Astronautics and Aeronautics, Vol. 163.

Parkvall S, Dahlman E, Furuskär A, Jading Y, Olsson M and Stefan Wänstedt KZ 2008 LTE-Advanced –evolving LTE towards IMT-Advanced. *Proceedings 68th IEEE Vehicular Technology Conference (VTC 2008–Fall)*, pp. 1–5, Calgary, Canada.

Paulraj A, Nabar R and Gore D 2003 *Introduction to Space-Time Wireless Communications*. Cambridge University Press, Cambridge, UK.

Perälä T and Ali-Löytty S 2008 Kalman-type positioning filters with floor plan information. *Proceedings of the International Conference on Advances in Mobile Computing & Multimedia (MoMM)* pp. 350–355.

Perälä T and Piché R 2007 Robust Extended Kalman Filtering in Hybrid Positioning Applications. *Proceedings of the Workshop on Positioning, Navigation and Communication (WPNC)* pp. 55–63.

Perkins C 1998 Mobile networking through mobile IP. *Internet Computing, IEEE* **2**(1), 58–69.

Perkins C and Jagannadh T 1995 DHCP for mobile networking with TCP/IP *Proceedings of the IEEE Symposium on Computers and Communications*, 1995, pp. 255–261.

Pichler M, Schwarzer S, Stelzer A and Vossiek M 2009 Multi-channel distance measurement with IEEE 802.15.4 (ZigBee) devices. *IEEE J. Sel. Top. Signal Process.* **3**(5), 845–859.

Pitt M and Shephard N 1999 Filtering via simulation: Auxiliary particle filters. *Journal of the American Statistical Association* **94**(446), 590–599.

Plass S 2008 *Cellular MC-CDMA Downlink Systems – Coordination, Cancellation, and Use of Inter-Cell Interference* PhD thesis Universität Ulm, Germany. VDI Verlag Düsseldorf, Series 10, No. 788, ISBN 978-3-18-378810-1.

POLE STAR 2013 Indoor positioning www.polestar.eu/en/nao-campus/indoor-positioning.html [Online; accessed November 18, 2013].

Prasithsangaree P, Krishnamurthy P and Chrysanthis P 2002 On indoor position location with wireless LANs. *Proceedings of the 13th IEEE International Symposium on Personal Indoor and Mobile Radio Communications (PIMRC 2002)*, vol. **2**, pp. 720–724, Lisbon, Portugal.

Press WH, Teukolsky SA, Vetterling WT and Flannery BP 1992 *Numerical recipes in C (2nd edn): the art of scientific computing*. Cambridge University Press, New York, NY, USA.

Prieto J, Bahillo A, Mazuelas S, Lorenzo R, Blas J and Fernandez P 2009 Adding indoor location capabilities to an IEEE 802.11 WLAN using real-time RTT measurements. *Wireless Telecommunications Symposium (WTS) 2009*, pp. 1–7.

Proakis JG 2001 *Digital Communications* 4th edn. McGraw-Hill.

Puscher F 2011 Gut oder Schein. *Magazin für Computer Technik (c't)* **2011**(7), 142–145.

Qi Y, Kobayashi H and Suda H 2006 Analysis of wireless geolocation in a non-line-of-sight environment. *IEEE Transactions on Wireless Communications* **5**(3), 672–681.

Qualcomm Europe 2009 On OTDOA in LTE. Technical Report R1-090353, 3GPP TSG-RAN WG1 #55bis, Ljubljana, Slovenia.

Riba J and Urruela A 2004 A non-line-of-sight mitigation technique based on ML-detection. *Proceedings of the IEEE International Conference on Acoustics, Speech, and Signal Processing (ICASSP)*.

Ristic B, Arulampalam S and Gordon N 2004 *Beyond the Kalman Filter–Particle Filters for Tracking Applications*. Artech House.

Roos T, Myllymäki P, Tirri H, Misikangas P and Sievänen J 2002 A probabilistic approach to WLAN user location estimation. *International Journal of Wireless Information Networks* **9**(3), 155–164.

Sand S, Wang W and Dammann A 2012 Cramér-Rao lower bounds for hybrid distance estimation schemes. *Proceedings of 76th IEEE Vehicular Technology Conference (VTC-Fall)*.

Schmidl TM and Cox DC 1997 Robust frequency and timing synchronization for OFDM. *IEEE Transactions on Communications* **45**(12), 1613–1621.

Schön T, Gustafsson F and Nordlund PJ 2005 Marginalized particle filters for mixed linear/nonlinear state-space models. *IEEE Transactions on Signal Processing* **53**(7), 2279–2289.

SIS 2002 ESA EGNOS Service 'Signal-in-Space through the Internet' (SISNeT). www.egnos-pro.esa.int/sisnet [Online; accessed November 18, 2013].

Smailagic A and Kogan D 2002 Location sensing and privacy in a context-aware computing environment. *IEEE Wireless Communications* **9**(5), 10–17.

Sternad M, Ottosson T, Ahlen A and Svensson A 2003 Attaining both coverage and high spectral efficiency with adaptive OFDM downlinks. *2003 IEEE 58th Vehicular Technology Conference (VTC 2003-Fall)*, vol. **4**, pp. 2486–2490.

Stüber GL 2001 *Principles of Mobile Communication* 2nd edn. Springer, Heidelberg, Germany.

Teldio 2013 Bluetooth indoor positioning module www.teldio.com/zonith/indoor_positioning [Online; accessed November 18, 2013].

Texas Instruments 2013 *Data Sheet Texas Instruments Chipcon CC2431 – System-on-Chip for 2.4 GHz ZigBee/IEEE 802.15.4 with Location Engine* Texas Instruments.

TimeDomain 2013 Pulson®410: The world's best ranging radio www.timedomain.com/p400.php [Online; accessed November 18, 2013].

Tsai Y, Zhang G, Grieca D, Ozluturk F and Wang X 2007 Cell search in 3GPP long term evolution systems. *IEEE Vehicular Technology Magazine* **2**(2), 23–29.

TSG-RAN W 2001 Status of UE positioning *TSG-RAN Meeting #12*, Stockholm, Sweden.

UbiSense 2013 UbiSense real-time location systems (RTLS) www.ubisense.net/en/rtls-solutions/ [Online; accessed November 18, 2013].

Uhlmann JK 1994 Simultaneous map building and localization for real time applications. Technical report, University of Oxford.

Urruela A, Sala J and Rima J 2006 Average Performance Analysis of Circular and Hyperbolic Geolocation. *IEEE Transactions on Vehicular Technology* **55**(1), 52–66.

Venkatachalam M, Etemad K, Ballantyne W and Chen B 2009 Location services in WiMAX networks. *IEEE Communications Magazine* **47**(10), 92–98.

Wallbaum M and Wasch T 2004 Markov localization of wireless local area network clients. *Proceedings of the 1st IFIP-TC6 Working Conference on Wireless-On-Demand Network Systems (WONS 2004)*, vol. 2928 of *Lecture Notes in Computer Science*, pp. 135–154, Madonna di Campiglio, Italy.

Wan EA and van der Merwe R 2000 The unscented Kalman filter for nonlinear estimation. *Proceedings of the Adaptive Systems for Signal Processing, Communications, and Control Symposium* pp. 153–158.

Wang F, Ghosh A, Sankaran C, Fleming PJ, Hsieh F and Benes SJ 2008a Mobile WiMAX systems: Performance and evolution. *IEEE Communications Magazine* **46**(10), 41–49.

Wang S, Lee KD, Kim SG and Kwak JS 2008b *Enhance Downlink Positioning in WiMAX/16m, document IEEE C802.16m-08/1106r3*. LG Electronics.

Wang W, Jost T, Mensing C and Dammann A 2009a TOA and TDOA error models for NLOS propagation based on outdoor to indoor channel measurement. *Proceedings of the IEEE Wireless Communications and Networking Conference (WCNC) 2009*, pp. 1–6, Budapest, Hungary.

Wang W, Jost T, Mensing C, Dammann A and Fawaz K 2009b Indoor propagation effects on TOA error for joint GNSS and terrestrial radio based localization. *Proceedings of the IEEE Vehicular Technology Conference (VTC) Spring*, pp. 1–5, Barcelona, Spain.

Wendlandt K, Berhig M and Robertson P 2005 Indoor localization with probability density functions based on Bluetooth *IEEE 16th International Symposium on Personal, Indoor and Mobile Radio Communications (PIMRC) 2005*, vol. **3**, pp. 2040–2044.

Wendlandt K, Sukchaya K, Robertson P, Khider M and Angermann M 2007 Demonstration of a realtime active-tag RFID, Java based indoor localization system using particle filtering. *Adjunct Proceedings of the 9th International Conference on Ubiquitous Computing (UBICOMP)*, pp. 108–111, Innsbruck, Austria.

WHE n.d. EU-ICT FP7 Project *Wireless Hybrid Enhanced Mobile Radio Estimators (WHERE)* www.ict-where.eu [Online; accessed November 18, 2013].

WHERE D2.1 2008 Performance assessment of hybrid data fusion and tracking algorithms. Technical report, Deliverable, Wireless Hybrid Enhanced Mobile Radio Estimators (WHERE) Project, ICT-217033.

WHERE D2.3 2010 Hybrid localization techniques. Technical report, Deliverable, Wireless Hybrid Enhanced Mobile Radio Estimators (WHERE) Project, ICT-217033.

WHERE D2.4 2010 Performance of WHERE cooperative positioning techniques.Technical report, Deliverable, Wireless Hybrid Enhanced Mobile Radio Estimators (WHERE) Project, ICT-217033.

WHERE D3.1 2009 Physical layer enhancements using localization data. Technical report, Deliverable, Wireless Hybrid Enhanced Mobile Radio Estimators (WHERE) Project, ICT-217033.

WHERE D3.4 2010 Location based optimisation for PHY algorithms/protocols. Technical report, Deliverable, Wireless Hybrid Enhanced Mobile Radio Estimators (WHERE) Project, ICT-217033.

WHERE D3.6 2010 Relaying and cooperative communications enhancements based on positioning data (final report). Technical report, Deliverable, Wireless Hybrid Enhanced Mobile Radio Estimators (WHERE) Project, ICT-217033.

WHERE D3.7 2010 Optimized cellular connectivity using positioning data. Technical report, Deliverable, Wireless Hybrid Enhanced Mobile Radio Estimators (WHERE) Project, ICT-217033.

WHERE D4.3 2008 Modelling of the channel and its variability. Technical report, Deliverable, Wireless Hybrid Enhanced Mobile Radio Estimators (WHERE) Project, ICT-217033.

Wikipedia 2013a Light cone – Wikipedia, the free encyclopedia. [Online; accessed November 18, 2013].

Wikipedia 2013b List of UMTS networks – Wikipedia, the free encyclopedia. [Online; accessed November 18, 2013].

WiMedia 2009a *Distrubted Medium Access Control (MAC) for Wireless Networks 1.5* WiMedia Alliance.

WiMedia 2009b *MultiBand OFDM Physical Layer Specification 1.5* WiMedia Alliance.

WiMedia 2013 WiMedia Alliance www.wimedia.org/en/index.asp [Online; accessed November 18, 2013].

WirelessHD 2010 *WirelessHD Specification Version 1.1 Overview* WirelessHD Consortium.

Wymeersch H, Lien J and Win M 2009 Cooperative localization in wireless networks. *Proceedings of the IEEE* **97**(2), 427–450.

Xiang Y, Luo J and Hartmann C 2007 Inter-cell interference mitigation through flexible reuse in OFDMA based communication networks. *European Wireless Conference, Paris*.

Xu H, Chong CC, Guvenc I, Watanabe F and Yang L 2008 High-resolution TOA estimation with Multi-Band OFDM UWB signals. *IEEE International Conference on Communications (ICC) 2008*, pp. 4191–4196.

Yuillel AL and Grzywacz NM 1988 The motion coherence theory *Proceedings Second International Conference on Computer Vision*, pp. 344–353.

ZEBRA 2013 Dart sensors www.zebra.com/gb/en/products-services/location-solutions/dart-uwb/dart-sensor.html [Online; accessed November 18, 2013].

Zekavat SAR and Buehrer RM 2012 *Handbook of Position Lcoation: Theory, Practice, and Advances*. John Wiley & Sons, Inc., Hoboken, NJ, USA.

Zhang L, Liu X, Song J, Gurrin C and Zhu Z 2013 A comprehensive study of Bluetooth fingerprinting-based algorithms for localization. 27th *International Conference on Advanced Information Networking and Applications Workshops (WAINA) 2013*, pp. 300–305.

Zhou S and Pollard J 2006 Position measurement using Bluetooth. *IEEE Transactions on Consumer Electronics* **52**(2), 555–558.

Zhu H and Cao G 2005 rDCF: A relay-enabled medium access control protocol for wireless ad hoc networks. *Proceedings IEEE INFOCOM 2005. 24th Annual Joint Conference of the IEEE Computer and Communications Societies*, vol. 1.

Zhu J, Zeng K, Kim KH and Mohapatra P 2012 Improving crowd-sourced Wi-Fi localization systems using Bluetooth beacons. *2012 9th Annual IEEE Communications Society Conference on Sensor, Mesh and Ad Hoc Communications and Networks (SECON)*, pp. 290–298.

ZigBee 2007 *ZigBee-2007 Specification*. ZigBee Alliance. version r17.

ZigBee Alliance 2013 ZigBee www.zigbee.org/Home.aspx [Online; accessed November 18, 2013].

Index

Positioning in Wireless Communications Systems, First Edition. Stephan Sand, Armin Dammann and Christian Mensing.
© 2014 John Wiley & Sons, Ltd. Published 2014 by John Wiley & Sons, Ltd.

Printed and bound by CPI Group (UK) Ltd, Croydon, CR0 4YY

27/10/2024

14580293-0002